III International Scientific and Technical Conference "Materials Technologies and Equipment for the Development of the Arctic and Siberia" (ARCTIC-2019)

IOP Conference Series: Materials Science and Engineering
Volume 696

Tomsk, Russia
25 - 28 September 2019

ISBN: 978-1-7138-0988-3
ISSN: 1757-8981

Printed from e-media with permission by:

Curran Associates, Inc.
57 Morehouse Lane
Red Hook, NY 12571

Some format issues inherent in the e-media version may also appear in this print version.

Institute of Physics
Dirac House, Temple Back
Bristol BS1 6BE UK

Phone: 44 1 17 929 7481
Fax: 44 1 17 920 0979

techtracking@iop.org

Additional copies of this publication are available from:

Curran Associates, Inc.
57 Morehouse Lane
Red Hook, NY 12571 USA
Phone: 845-758-0400
Fax: 845-758-2633
Email: curran@proceedings.com
Web: www.proceedings.com

TABLE OF CONTENTS

IOP Conf. Series: Materials Science and Engineering **696** (2019) 011001 doi:10.1088/1757-899X/696/1/011001

Preface

The III International Conference 'Materials, Technology and Equipment for the development of the Arctic and Siberia' (ARCTIC-2019) was held by Tomsk State University (Tomsk) and the All-Russian Scientific Research Institute of Aviation Materials-VIAM (Moscow) on September 25–28, 2019 in Tomsk.

The Conference goal was to intensify research in the fields of special materials, technology and facilities used in the development of the Arctic and Siberia, increasing the effectiveness of the implementation of the scientific results, strengthening the cooperation between industry and famous scientists of different countries.

The conference covers the fundamental and applied problems of the scientific and technical support for the development of the Arctic Zone of Russian Federation, as well as the current state and prospects of the Arctic and Siberia development.

The mains topics of the conference were as follows:

- *Arctic materials (research, producing, application)*
- *Technology of extraction and transportation of raw materials of the Arctic and Siberian zones*
- *Technology for local processing of Arctic raw materials*
- *Peculiar properties of technical devices and facilities for the territories with harsh climate conditions*
- *Testing materials and devices in harsh climate conditions and the development of scientific and technical documentation*
- *Environmental factors for the development of the Arctic and Siberia*

These proceedings papers reflect the current fundamental and applied research in areas of special materials, technologies, equipment and structures used in the development of the Arctic and Siberia, improving the effectiveness of the application of research results and their practical use.

The Organizing and Program Committees of ARCTIC-2019

IOP Conf. Series: Materials Science and Engineering **696** (2019) 011002 doi:10.1088/1757-899X/696/1/011002

Peer review statement

All papers published in this volume of *IOP Conference Series: Materials Science and Engineering* have been peer reviewed through processes administered by the proceedings Editors. Reviews were conducted by expert referees to the professional and scientific standards expected of a proceedings journal published by IOP Publishing.

IOP Conf. Series: Materials Science and Engineering **696** (2019) 012001 doi:10.1088/1757-899X/696/1/012001

Enhanced Oil Recovery Technologies for Arctic and Siberian Regions

L K Altunina[1, 2], V A Kuvshinov[1], I V Kuvshinov[1], L A Stasieva[1]

[1]Institute of Petroleum Chemistry, Siberian Branch of the Russian Academy of Sciences (IPC SB RAS), 4, Akademichesky Avenue, 634055, Tomsk, Russia
[2]National Research Tomsk State University, 634050 Tomsk, Lenin Avenue, 36

E-mail: alk@ipc.tsc.ru

Abstract. The results of laboratory research, field testing and industrial use of physicochemical and integrated enhanced oil recovery (EOR) technologies for fields of Arctic and Siberian regions are presented. The EOR technologies are based on the concept of using reservoir energy to generate gels, sols and surfactant systems preserving in the reservoir a complex of properties that are optimal for the purposes of oil displacement. The technologies using gel-forming compositions and surfactant-based oil-displacing systems are applicable in extreme climatic conditions, effective in a wide range of geological and physical conditions and stages of field development, cost-effective and environmentally friendly.

1. Introduction

Currently, the development of the natural resources of the Arctic is a priority state task. Its solution is complicated by extreme climatic conditions, which place high demands on the technologies and materials used. In the coming decades, the Arctic region in Russia will be the main reserve of the country's oil and gas industry. For the effective development of oil and gas fields in the Arctic, it is necessary to create and apply large-scale scientifically sound technologies for oil and gas production adapted to northern conditions, to develop new chemicals for implementation of technologies [1-4].

For more than 30 years, IPC SB RAS has been carrying out fundamental and applied studies of natural hydrocarbon systems, on the basis of which new technologies for oil production are created. To intensify the development and improve oil recovery of oil and gas facilities with difficult-to-recover reserves, including deposits of heavy, highly viscous oils, a promising concept has been developed to use the energy of the formation or injected heat carrier to generate chemical "smart" nanoscale systems: gels, sols, solutions of surface-active substances (surfactants) and buffer systems with adjustable alkalinity, preserving, self-sustaining in the formation for a long time a complex of colloidal-chemical properties, optimal for oil displacement [4, 5]. Eleven EOR technologies have been created that are industrially used by the oil companies LUKOIL, ROSNEFT, etc., and 200-300 wells are processed annually. Through their use over the past 5 years, more than 3 million tons of oil have been additionally produced. Industrial production of a number of compositions has been organized.

The accumulated knowledge and scientific and technical experience of the IPC SB RAS were used to adapt the developed technologies and create new technologies for increasing oil recovery using gel-forming and oil-displacing systems for Arctic and Siberian regions, which are applicable in extreme climatic conditions, effective in wide range of geological and physical conditions and stages of field development feasible using standard oilfield equipment. Chemical reagents for performing technologies are environmentally safe products of large-tonnage commercial production. They have solid commodity form or low freezing point (minus 20 ÷ minus 60 °C), and can be delivered by any transport.

2. Oil-displacing systems based on surfactants and buffer systems

Promising for oil fields in Siberian and Arctic regions are technologies for increasing oil recovery, reducing water cut in production wells and intensifying oil production using smart oil-based surfactant compositions with adjustable viscosity and alkalinity: NINKA®, thickened NINKA-Z, GBK and IHN-PRO. The technologies are applicable in difficult operating conditions, including for highly viscous oil fields developed both with the use of thermal methods and without thermal-steam stimulation [4-6].

Since 2003, on the Permian-Carboniferous reservoir of the high-viscosity oil in the Usinsk field, LUKOIL-Komi LLC, together with the IPC SB RAS, LUKOIL-Engineering LLC and OSK LLC, have been conducting field tests of integrated technologies of steam-thermal and physicochemical treatment and industrial use of the developed technologies, as well as "cold" physicochemical technologies to inhance oil recovery and intensify heavy oil production under natural conditions, without heat exposure, using gel-forming and oil-displacing systems [4-6].

Oil-displacing compositions based on the surfactant – carbamide – ammonium salt – water system have been created at the IPC SB RAS, which in the formation at high temperature or due to the thermal energy of the injected heat carrier generate carbon dioxide CO_2 and an alkaline buffer system (NINKA® compositions), which contribute to a decrease in oil viscosity and additional its displacement [4, 5]. In the reservoir, under the influence of high temperature, carbamide is hydrolyzed to form CO_2 and ammonia NH_3. CO_2, unlike ammonia, is much more soluble in oil than in water. Therefore, in the oil – water system, the oil phase will be enriched in CO_2, and the aqueous phase will be enriched in NH_3, which forms an alkaline system with an ammonium salt with a maximum buffer capacity in the pH range of 9 ÷ 10 [5], which is optimal for oil displacement. The dissolution of CO_2 in oil leads to a decrease in its viscosity, Figure 1. CO_2 and NH_3 reduce the swelling of clay minerals of the reservoir rock and thereby contribute to maintaining the initial permeability of the formation. The ammonia buffer system formed during the dissolution of ammonia in an aqueous solution of ammonium salts plays the same role. In addition, due to its alkalinity, pH 9÷10, and the presence of surfactants, it contributes to the additional displacement of oil, reduction of interfacial tension and degradation, thinning of highly viscous layers or films at the oil – water – rock interfaces, which worsen the filtration of liquids in the reservoir and reduce the completeness of oil production.

On the Permian-Carboniferous reservoir of the Usinsk oil field and on the Gaoshen oil field (China) in 2003-2007 field tests of NINKA® systems were conducted. It was found that their use in stationary steam injection leads to a decrease in water cut by 10-20 % and an increase in oil production by an average of 40 %. Under the cycle-steam stimulation (CSS) with NINKA® systems, an increase in oil production is observed by 1.5-3 times, a decrease in oil viscosity by 2-3 times, Figure 2. Currently, technologies using NINKA® systems are used on an industrial scale, for example, Figure 2.

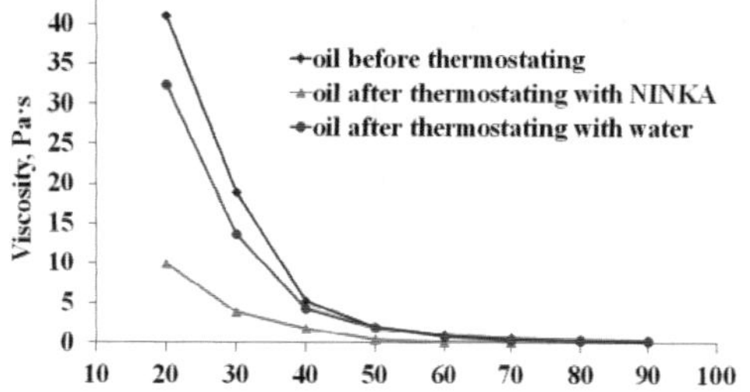

Figure 1. Decreasing oil viscosity of Liaohe field after thermostating during 6 h at 200 °C with NINKA® system.

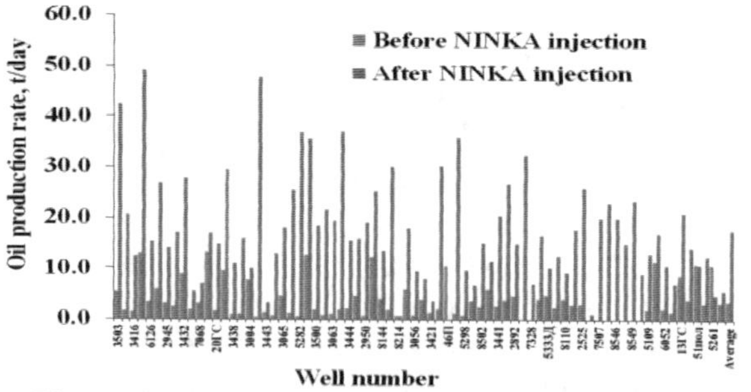

Figure 2. Commercial use of EOR technology in 2017, increase in oil production rate for 37 production wells in Usinsk oil field after the CSS with NINKA® system.

A further development of the work was the creation of compositions, suitable for the Arctic and Siberian regions, with adjustable viscosity and alkalinity, which increase of both the oil displacement coefficient and the coverage coefficient of the formation during water and steam flooding. A thickened oil-displacing composition based on surfactants, ammonium salts, aluminum and carbamide (NINKA-Z) was created, which, as a result of chemical evolution directly in the formation, becomes both a flow-rejecting and oil-displacing composition [4]. As mentioned above, in the formation, at high temperature carbamide is hydrolyzed, forming CO_2 and NH_3, which with an ammonium salt gives an alkaline buffer system. Raising the pH causes hydrolysis of the Al salt to form Al hydroxide sol, while the viscosity of composition increases by 1-2 orders of magnitude. This leads to increase in thermal coverage of the formation, the inclusion of low-permeability layers, decrease in the viscosity of the oil and its washing out. As a result, there is an increase in the coefficient of coverage of the reservoir, an increase in the oil recovery factor (ORF) and the intensification of oil production.

Studies of the kinetics of sol formation and the rheological properties of solutions and sols obtained at 90, 150, and 200 °C showed that after thermostating of sol-forming oil-displacing composition with adjustable viscosity and alkalinity, depending on the concentration of aluminum salt, the viscosity of the composition increases 6-78 times, the pH of the composition after thermostating rises to 7.7-10.1 u. As an example, Figure 3 shows the results of a study of the rheological properties of the composition before and after the formation of sols due to thermostating at 150 °C for 5 h. The measurements were carried out after cooling the composition to 20 °C. Before thermostating, the composition is a Newtonian fluid, after the formation of the sol, it is a viscoplastic fluid, Figure 3.

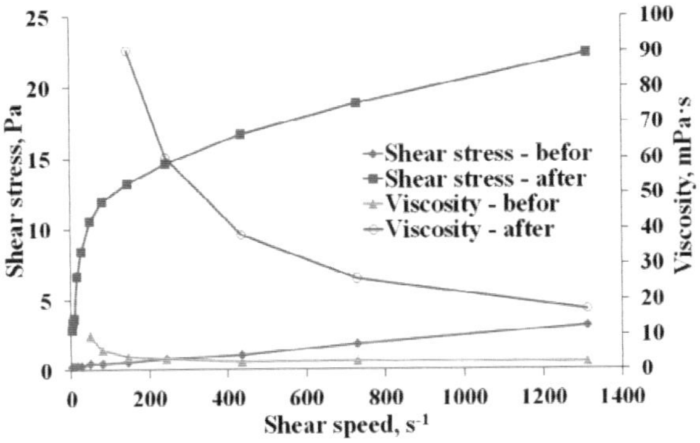

Figure 3. The complete rheological flow curves and dependence of NINKA-Z system viscosity before and after 5 h of thermostating at 150 °C: before thermostating system is Newtonian fluid, after the sol formation – viscoplastic fluid.

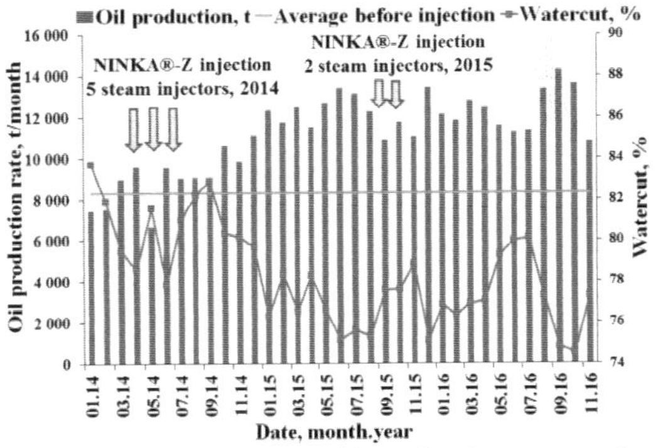

Figure 4. Increase in oil production rate and decrease in water cut after injection of the NINKA-Z system in 2014-2015 into 7 steam-injection wells on the pilot site of the PTV-South-West in the Usinsk oil field.

In 2014-2015 field tests of the composition were carried out, the NINKA-Z thickened composition was pumped into 7 steam injection wells in the pilot site of the steam stimulation on the Permian-Carboniferous reservoir of the Usinsk oil field, Komi Republic (Figure 4). The injection volume was 80-110 m^3 per well, the effect was monitored for 75 producing wells of the site. According to the results presented in Figure 4, a steady decrease in water cut and an increase in oil production after injection are seen. The total effect on the site amounted of more than 70 thousand tons of additional oil. The technology is recommended for commercial application.

To enhance oil recovery and intensify the development of high-viscosity oil deposits in the absence of steam stimulation, at 20-40 °C, oil-displacing compositions using surfactants and polyols with controlled viscosity were developed at the IPC SB RAS: alkaline composition IKhN-PRO, based on inorganic buffer system, and acid composition GBK of prolonged action, based on inorganic acid adduct [4, 6]. Compositions have a freezing temperature of minus 20 ÷ minus 60 °C, low interfacial tension at the oil boundary, adjustable density, from 1100 to 1300 kg/m^3, and viscosity – from tens to hundreds of mPa·s, applicable in the natural mode of development of high-viscosity oil deposits. The GBK composition is most effective in carbonate reservoirs, has a delayed reaction with carbonate rocks, does not form insoluble products, and restores the initial permeability of the reservoir.

In 2014, to increase the production rate of low-productive producing wells of the Permian-Carboniferous reservoir of the Usinsk oil field without steam heat stimulation, the IKhN-PRO composition was injected into 5 and GBK into 10 production wells, Figure 5.

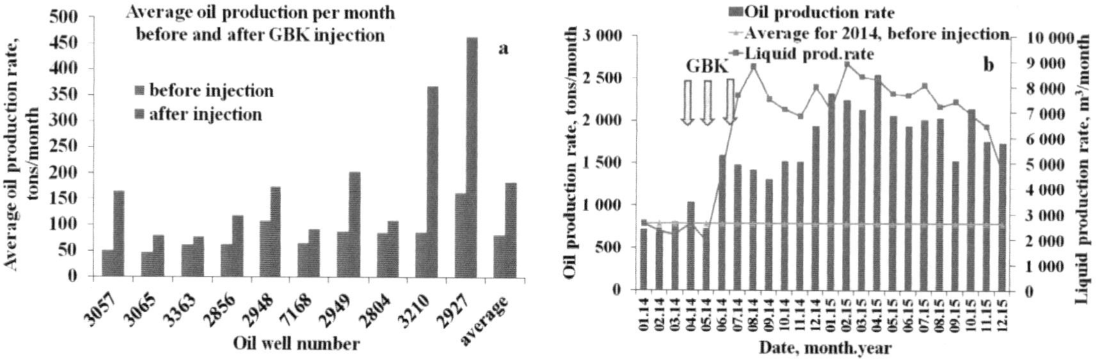

Figure 5. The results of pilot tests in low-yield production wells of the Permian-Carboniferous reservoir of the Usinsk oil field using acid oil-displacing system (as a result of 19 months): a – the increase of the average values of the monthly oil production rate before and after injection; b – increase in oil production and liquid flow rates total for 10 wells.

After injection of the compositions, the oil production rate increased by 5.5-14.8 tons/days and liquid flow rates – by 15-25 m^3/day. Incremental oil production amounted to 28 thou tons, 1600-2000 tons/well. The EOR technologies were recommended for commercial application. All components of the systems are environmentally safe products of large-tonnage industrial production. The technologies are applicable for the Arctic and Siberian regions.

3. Thermotropic gel-forming systems for enhanced oil recovery and for water shutoff

To enhance/improve oil recovery (EOR/IOR) and limit water inflow to deposits with highly heterogeneous reservoirs, including those in the northern and arctic regions, gel technologies have been created at the IPC SB RAS. They use thermotropic gel-forming systems, which under surface conditions are low-viscosity aqueous solutions, in reservoir ones they turn into gels. Gelation occurs under the influence of thermal energy of the reservoir or the injected heat carrier, without crosslinking agents [4, 5]. The gelation kinetics, rheological and filtration characteristics of various types of gels for heterogeneous formations with permeability in the range from 0.01 to 10 μm^2 were studied. Thermotropic gel-forming systems are proposed: polymer based on cellulose ethers (CE) and carbamide (METKA®) and inorganic systems "aluminum salt – carbamide – water" (GALKA®) with different time of gelation – from several minutes to several days – in the range temperatures 30-320 °C. Using them, eight gel technologies have been developed to improve oil recovery in highly heterogeneous formations that are industrially used in fields in Western Siberia and Komi Republic [4-6]. The environmental safety of reagents and their safety for humans make it possible to widely use gel technology in the fields of Russia and other countries.

The gel-forming ability of thermotropic systems was studied by measuring the viscosity and shear stress using Haake RheoStress 600 rotational viscometer at a temperature of 20 to 150 °C and a pressure of up to 50 atm under dynamic conditions (with full compression). At temperature of 100-150 °C, after a certain time, the solution turns into a solid-like gel of a coagulation structure with pronounced thixotropy and yield strength of 25-90 Pa, Figures 6, 7. The yield strength increases with increasing loading rate. The type of rheological dependences indicates the viscoelastic properties of the gels. It should be noted that in the shear rates range of 0.01-5 s^{-1} at high pressures, the shear stress and gel viscosity in the systems: aluminum salt – carbamide – water and CE – carbamide – water are several orders of magnitude higher, figures 6, 7, than at atmospheric pressure [5]. The addition of carbamide and ammonium rhodanide to CE solution leads to increase in gelation temperature, while the effect of the additives is additive: when the carbamide and ammonium rhodanide are combined, the gelation temperature can be higher than 100 °C (Figure 7). The gels are stable and retain their rheological properties at high temperatures – up to 150-200 °C.

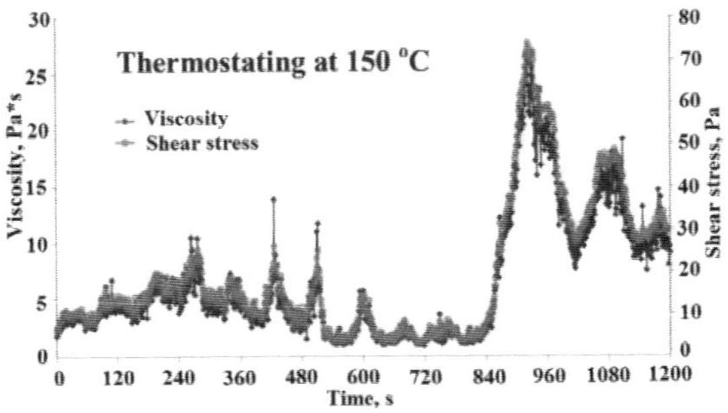

Figure 6. Change in viscosity and shear stress with time for the GALKA®-S system at shear rate of 3 s^{-1}.

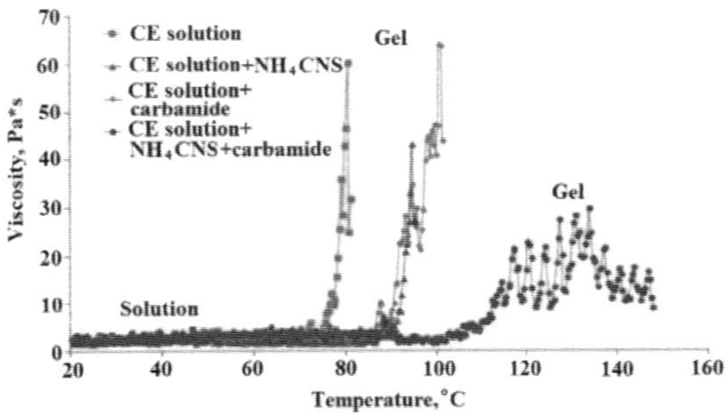

Figure 7. Change in viscosity during the solution – gel phase transition for the METKA® system at shear rate of 3 s^{-1}.

Technologies for Arctic and Siberian fields are promising using thermotropic gel-forming systems: inorganic (GALKA®), polymer (METKA®) and systems with two gel-forming components – polymer and inorganic (MEGA), which directly in the reservoir form cohesive nanosized "gel-in-gel" structures with improved rheological characteristics and structural-mechanical properties [4-6].

To improve the effectiveness of the thermal-steam stimulation due to selective water shutoff, in 2014 the METKA® system was injected into 5 production wells in the Usinsk oil field, on the site of the areal steam injection. The volume of the injected system was 19-95 m^3 per well. After the injection of the METKA® system one observed the increase in the oil production rate and the decrease in the water cutting of the production wells, Figure 8, a. Incremental oil production was ~ 11,000 tons, or 2,100 tons per well. The treatment was effective for 16 months. Based on the results of the pilot tests, the EOR technology has been recommended for commercial application.

For water shutoff and to enhance oil recovery at reservoir temperatures of 60-220 °C, at water flooding, thermal-steam and cycling-steam stimulations IPC SB RAS has created a high-temperature gel-forming MEGA system with two gel-forming components – polymer and inorganic, based on the "aluminum salt – cellulose ether – carbamide – water". The system generates *in situ* connected-disperse nano-sized structures of the "gel-in-gel" type with improved rheological characteristics and structurally mechanical properties [4, 6]. The reservoir heat energy or that of the injected heat carrier without a cross-linking agent is a factor causing gellation. When heated above the lower critical temperature of dissolving the cellulose ether, in the system due to a phase transition first a polymer gel is formed and then inside the polymer gel an aluminum hydroxide gel is formed by a mechanism of hydrolytic polycondensation initiated by the products of the carbamide hydrolysis. As a result the structural and mechanical properties of the gel are improved, its viscosity and elasticity multiply increase. The gels formed in reservoir restrain water or steam breakthroughs from injection to production wells and redistribute the filtration fluid flows in reservoir. It stabilizes or decreases water cuttings of the surrounding production or steam-cyclic wells and thereby increases oil production.

The first pilot tests of the EOR technology employing the gel-forming nano-structured MEGA system for water shutoff and to enhance oil recovery were carried out at the end of 2016 at five production wells on the Permian-Carboniferous reservoir in Usinsk field at CSS and on the site of the areal steam injection. The volume of the injected system amounted to 80-120 m^3 per well. After well treatment one observed significant decrease in water cut, by 12-40 %, and multiple increase in the oil production rates, Figure 8, b. The pilot test results proved the ability of the MEGA system to block effectively the water inflow into the production well.

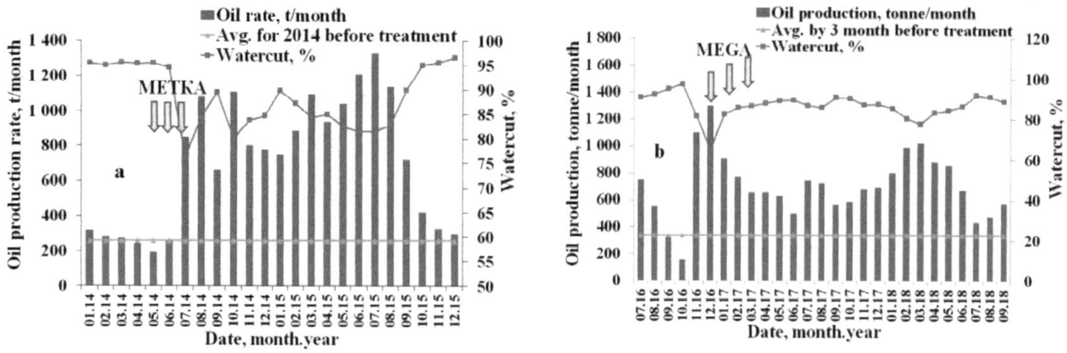

Figure 8. The results of pilot tests in the water shutoff area using METKA® (a) and MEGA (b) systems total for 5 production wells in Permian-Carboniferous reservoir of the Usinsk oil field: increased oil production rate and reduced water cut.

In 2015-2018 more than 190 wells have been stimulated on the Permian-Carboniferous reservoir of high-viscosity oil in the Usinsk oil field using the EOR/IOR technologies created at the IPC SB RAS. The additional oil production amounted to more than 900 thousand tons. The use of gels with areal steam stimulation leads to an increase in the coverage of the formation by steam injection, a decrease in water cut by 3-45 %, an increase in oil production by 11-33 %, and a decrease in liquid production by 14-25 %. Under the CSS, the increase in oil production was from 3 to 24 tons per day per well, and additional oil production averaged ≈1,000 tons per well treatment.

4. Conclusions

At the Institute of Petroleum Chemistry SB RAS a promising concept has been developed for using the energy of the formation or injected heat carrier to generate chemical "smart" nanoscale systems directly in the reservoir: gels, sols, surfactant solutions and buffer systems with controlled alkalinity, which preserve and maintain a complex of colloidal chemical properties in the reservoir for a long time optimal for oil displacement. With their application, new physicochemical and integrated technologies have been created to enhanced/improved oil recovery and intensify the development of deposits in the Arctic and Siberian regions, including deposits of highly viscous oils.

The EOR/IOR technologies are applicable in extreme climatic conditions, effective in a wide range of geological and physical conditions and stages of field development, and are carried out using the applied oilfield equipment. Chemical reagents for the implementation of EOR/IOR technologies are environmentally friendly products of large-tonnage industrial production, have a solid commodity form or low freezing temperature (minus 20 ÷ minus 60 °C), are delivered by any means of transport.

The industrial application of new created EOR/IOR technologies will allow for the profitable exploitation of fields in the Arctic and Siberia, and will contribute to the development of the oil industry in these regions.

Acknowledgement

The work was financially supported by the Ministry of Science and Higher Education of the RF, within the framework of the Project V.46.2.3 (No. 0370-2019-0002)".

References

[1] Donga X, Liua H, Chen Z, Wua K, Lua N, Zhanga Q 2019 *Applied Energy* **239** 1190–1211
[2] Zolotukhin A, Gudmestad O, Yarlsbyu E 2011 *Resources of oil and gas, development of offshore deposits* (Southampton, Great Britain, WIT press) 279 pp
[3] Wang Y, Hou J, Song Z, Yuan D, Zhang J, Zhao T 2018 *SPE Reservoir Evaluation& Engineering* **21**(01) 109–121
[4] Altunina L, Kuvshinov V, Stasieva L, Kuvshinov I 2018 *Chemistry for Sustainable Development* **26** (3) 261–277
[5] Altunina L and Kuvshinov V 2007 *Rus. Chem. Rev.* **76** (10) 1034–1052

[6] Altunina L, Kuvshinov V, Kuvshinov I, Stasieva L, Chertenkov M, Shkrabyuk L and Andreev D 2017 *Oil industry* **7** 26–29

IOP Conf. Series: Materials Science and Engineering **696** (2019) 012002 doi:10.1088/1757-899X/696/1/012002

The influence of seasonal changes on the formation of colloids and thin films of iron(III) hydroxide in Ob floodplain lakes

E Lyutova, L Borilo, E Izosimova, L Kolesnichenko, I Rozhkova-Timina and L Spivakova

National Research Tomsk State University, 36 Lenina Avenue, Tomsk 634050, Russian Federation

e-mail: lyutova.tsu@mail.ru

Abstract: The research was conducted in the middle Ob floodplain. Several parameters (dissolved CO_2, pH, conductivity, temperature, humin substances, Fe) were measured in six floodplain lakes in summer and autumn periods 2018. Temperature, conductivity and dissolved CO_2 demonstrated the significant difference between these two samples.

1. Introduction

Floodplain is a territory, which is covered by water during the spring flood [1]. Suspended matter, dissolved chemical elements and gases move around the flood zone. The river spreads wide during the spring flood and provides the water exchange with the lakes. So the floodplain lakes influence huge territories [1].

Dissolved gases are very important in the floodplain water objects. Without oxygen (O_2) and carbon dioxide (CO_2) the life is impossible [2, 3]. pH shows the changes in alkalinity of water environment. The organic composition of natural waters is formed by substances entering water bodies as a result of natural biological processes and the interactions of these substances with surface slope and industrial effluents. Iron (Fe) plays an important role in geochemical processes, such as adsorption, desorption, and redox reactions in soils, sediments, and water. Biological activity of iron impacts the phytoplankton development and microbial population in the water body [4-7]. The natural aquatic environment is a complex system consisting of a dissolved and colloidal fraction, solid particles. Natural aqueous nanocolloids are usually defined as natural nanoparticles with a size from 1 nm to 1 μm [8, 9]. Natural colloidal materials play a decisive role as a mediator in biogeochemical processes, such as cycling of C, N and P, aggregation and sedimentation, as well as the fate of pollutants, as these colloids have a large surface area, high concentration and dense surface binding sites, contributing to strong interactions [10-12]. In addition to colloids formed in water layers when deep or soil waters are mixed, a significant part of the iron can pass into thin films on the surface of temporary puddles and small streams.

The aim of the work is to study the influence of seasonal factors on the deposition of dissolved organic carbon (DOC) with colloidal and thin-film iron(III) hydroxide in the floodplain of the Ob River and the synthesis of thin films.

2. Materials and methods

2.1. Study site

The researches were conducted in the Ob's floodplain. The river Ob is 3650 km long and goes through Siberia from south (Altai mountains) to the north (the Gulf of Ob). Ob's floodplain is tens km wide and is the second floodplain in the world. Ob is the long river and in dependence on geomorphological, hydrological and climatic conditions can be divided into three parts: the lower Ob (from Altai mountains to Tom mouths), the middle Ob (from Tom mouth to Irtysh mouth), and the upper Ob (from Irtysh mouth to Gulf of Ob) [13, 14]. Our researches were conducted in the floodplain lakes in the middle course of Ob, Tomsk region, the station Kaibassovo; N57.246142°, E84.181919° (Figure 1). This part of Ob is ice-covered six months per year; the average thickness of ice before the ice-breaking in April is about 90 cm [15].

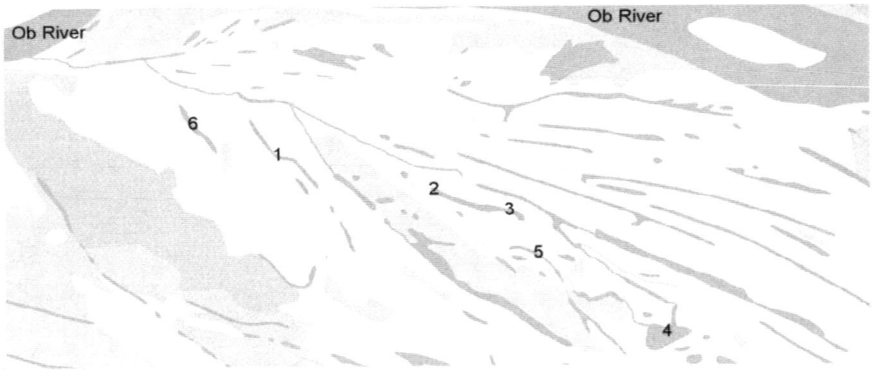

Figure 1. Study location map. The sampled lakes numbered from 1 to 6.

2.2. Data sampling and analysis

We measured and sampled six floodplain lakes in the summer and autumn 2018. We measured dissolved CO_2 in water using a data-logger with an underwater sensor (GM70 Hand-Held Carbon Dioxide Meter, Vaisala®). We put the logger in water at the depth 15 cm and waited for the reach of the curve's saturation on the indicator before noticing the measured value. Conductivity, pH, and temperature were measured by WTW Multi 3320 with data-loggers: pH-Electrode Sen Tix® 41 and WTW Tetra Con 325. The loggers were put in the water, and then we fixed the value.

The water samples were filtered in pre-washed 30-ml polypropylene Nalgene® flagons by Minisart single-use filter units (0.45 μm pore size, Sartorius, acetate cellulose filter). We analysed filtered water samples with a spectrometer in a 1-cm quartz cuvette (Eppendorf BioSpectrometer®).

To determine the concentration of organic acids, total content of iron and iron(III) used atomic absorption spectrometer 3300 manufactured by Perkin-Elmer. To obtain the coloured complexes by the spectrophotometric determination of the total content of iron and iron(III) we used sulfosalicylic acid (chemically pure), ammonia solution (chemically pure) and ammonium chloride (chemically pure). After the color emergence we measured the optical density of the solutions. The intensity of the color depends on the iron content and follows the Bouguer-Lambert-Beer law, if iron content in the analyzed water is up to 2 mg/ml. The measurements were carried out at a wavelength of 425 nm. Cuvettes have a working layer thickness of 2-5 cm. The calibration curve was constructed using iron ammonium alum with different concentrations at the same wavelength. According to the calibration graph, the iron content in the source water was calculated. To determine the content of organic acids measurements were carried out at a wavelength of 440 nm. To construct the calibration graph we used standardized samples of organic acids provided by the Siberian Research Institute of Agriculture and Peat.

IR spectra of dried solutions were analyzed on Agilent Cary 630 FTIR IR spectrometer in the frequency range of 400…4000 cm^{-1}.

Obtained data were analyzed using the software Statistica 6.0 and Excel 2010. The non-parametric Mann-Whitney U test was performed; a significant difference was set at $P < 0.05$.

For experimental (laboratory) modeling of the formation of colloids and thin films upon mixing of deep reduced waters and surface waters rich in dissolved organic carbon based on in situ measurements were selected iron, silicon, and humic organic compounds.

As raw components were taken $FeSO_4 \cdot 7H_2O$ (Russia) with a concentration of 2.5 mg/ml, dissolved organic matter isolated from peat bogs in the Tomsk Region, provided by the Siberian Research Institute of Agriculture and Peat (Tomsk) with a concentration in the range from 0.02 to 4.0 mg/ml, $Na_2SiO_3 \cdot 9H_2O$ (Russia) with a concentration in the range from 5 to 20 mg/ml.

For the synthesis of model solutions, combinations of components were selected: iron compounds and dissolved organic substances (sample 1), iron compounds and silicon compounds (sample 2), iron compounds, dissolved organic substances and silicon compounds (sample 3).

The stability of the synthesized solutions was evaluated by changing the optical density of the solution, changing the concentration of colloidal iron in the solution and particle size.

For a system of combining iron, dissolved organic substances and silicon compounds (sample 3), the coating thickness was determined using the method of ellipsometry on a laser ellipsometer SE400advanced.

3. Results and discussion

Lake water contains different dissolved elements and gases. The carbon dioxide is the product of the organisms' life activity. Iron (Fe) is one of the most abundant metals in freshwater. The chemical form of Fe (its chemical speciation) has a great influence on the mobility and availability of the element itself, as well as other nutrients and potentially toxic trace elements. Dissolved iron may have ionic form (the form of complex compounds with mineral and organic substances). In summer and autumn baseflow periods, the lake feeding is mostly groundwater, so the values of iron concentration remain almost unchanged (figure 2 a, b). However, for the lake 4, the summer and autumn values of iron differ 1.5 times. The change in iron concentration in the Ob River for the summer and autumn period remains constant at 0.7 mg/ml.

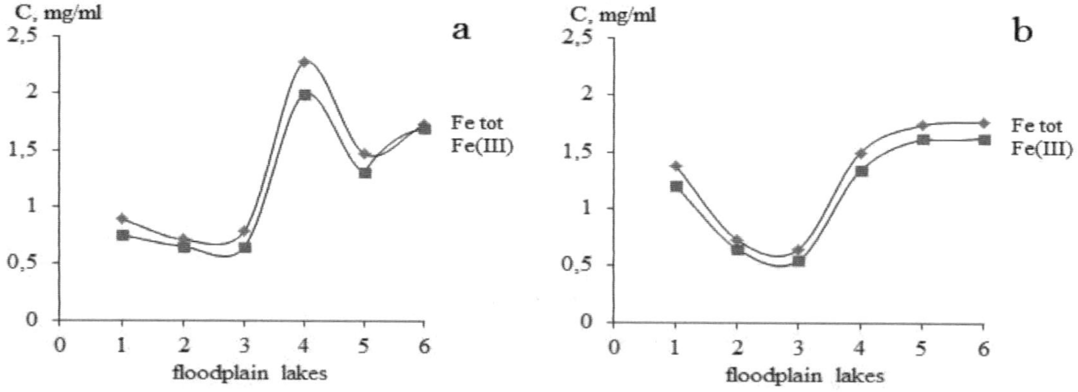

Figure 2. A graph of the concentration of total iron and iron(III); a – summer period; b – autumn period.

The values of CO_2 in two seasons differ 6.7 times. Dissolved CO_2 is stable in different floodplain lakes during the summer and during the autumn (figure 3a). Figure 3b shows that for all the studied floodplain lakes, the pH in the autumn period is in the range of 7.34–8.08, and in the summer it is slightly lower, but within the slightly alkaline region from 7 to 7.4. As is known [9], in an alkaline and slightly alkaline medium in water there is a large amount of iron(III) bound in complexes with humic acids. The concentration of organic acids in the summer-autumn period remains practically unchanged and ranges from 0.1 to 0.2 mg/ml.

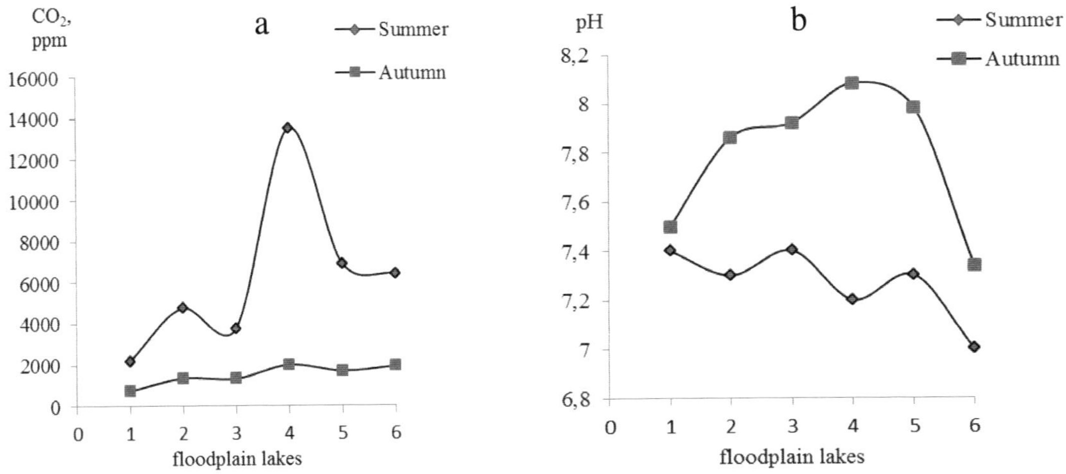

Figure 3. A graph of pH and CO_2 concentration in floodplain lakes; a – CO_2, b - pH.

We compared summer and autumn values of all measured parameters. We obtained the following results (table 1) using the descriptive statistics and non-parametric comparative Mann-Whitney test.

Table 1. Biogeochemical parameters and their significance by P<0.05.

Parameter	Mean ± st.dev. Summer	Mean ± st.dev. Autumn	Criteria U	Significance level p
Temperature t, C	23.51 ± 1.14	3.62 ± 0.48	0	**0.002700**
pH	7.27 ± 0.11	7.78 ± 0.24	2	**0.006494**
Conductivity,	145.18 ± 40.38	173.83 ± 67.78	13	0.667806
Dissolved CO_2, ppm	6258.33 ± 2688.33	1498.57 ± 365	0	**0.002700**
Humin substances, mg/l	0.018 ± 0.003	0.015 ± 0.005	12	0.150786
Fe(III), mg/l	1.17 ± 0.49	1.165 ± 0.38	18	0.773885
Fe total, mg/l	1.31 ± 0.51	1.295 ± 0.4	16	1

Table 1 shows that only temperature, pH and dissolved CO_2 showed the significant difference. The significant difference of temperature values in different seasons are natural because of climate peculiarities of Siberia region.

Our previous research [4, 16] demonstrated the negative correlation between dissolved oxygen and carbon dioxide in floodplain water. So we can suppose that dissolved oxygen significantly differs at different seasons. At summer the living activity is more intensive and the living organisms produce great quantities of CO_2. At autumn the flora and fauna start to die off and get to hibernation, but the photosyntetic rate is still high.

The forms of iron and their content in water are greatly influenced by pH, the concentration of organic acids (humic acids), dissolved oxygen and carbon dioxide, microorganisms that oxidize and reduce iron. In oxygen-rich water, breaking of the bonds of iron with organic matter is accompanied by the formation of Fe(III). In the absence of big amounts of organic matter, Fe(II) is rapidly oxidized with dissolved oxygen to Fe(III), and can precipitate as iron(III) hydroxides. The pH for the $Fe(OH)_3{\bullet}nH_2O$ deposition should be in the range 2.5-4.5, which is much lower than pH observed in the floodplain lakes. The probability of precipitation of iron hydroxides is low. Figure 3b shows that for all the floodplain lakes the pH in the autumn period is in the range of 7.34-8.08, and in the summer period it is slightly lower (from 7 to 7.4), but still within the weakly alkaline range. It is well known [1, 4, 9] that in the alkaline and weakly alkaline ranges in water there is a large amount of iron(III) bound in complexes with humic acids. On the other hand, the migration of iron in surface waters largely depends on the activity of microorganisms. Biological transformation of iron is carried out with the participation of groups of iron-reducing and oxidizing microorganisms, and the rate of biogenic processes of iron oxidation is many times higher than chemical oxidation. It has been proven [10-12], Fe is characterized by the highest affinity of all metals for complex formation with organic matter. Our research demonstrated that the concentration of organic acids in the summer-autumn period does not change. Thus, the concentration of humic acids, pH, CO_2 are the determining factors of the water geochemistry of iron.

The work carried out experimental (laboratory) modeling of the formation of colloids and thin films upon mixing of deep water and surface water rich in dissolved organic carbon. The resulting colloidal solutions are resistant to the sedimentation process for 30 days.

The formation of colloidal iron compounds in the presence of silicon compounds occurs in the entire studied range of silicon concentrations. At low concentrations of silicon ions, partial coagulation of iron compounds occurs, as evidenced by an increase in particle size. At low concentrations of silicon ions of 5.0 and 10.0 mg/ml, the average particle size of the dispersed phase is 180 nm. With an increase in the concentration of silicon ions to 20.0 mg/l, the particle size decreases to 70 nm. In the solution (composition 3), the concentration of dissolved organic matter is 4 mg/ml, the concentration of iron ions is 2.5 mg/ml. For the formation of a stable colloidal system, a concentration of silicon ion of 10 mg/ml is sufficient.

In addition to colloids formed in the water column, a substantial part of the iron can pass into thin films on the surface. Therefore, thin films on a silicon substrate (model surface) were obtained from model solutions. To establish the uniformity of the formation of the film coating, the surface of the film was evaluated using a scanning electron microscope. Figure 4 shows a micrograph of the film surface, which is uniform over the entire surface.

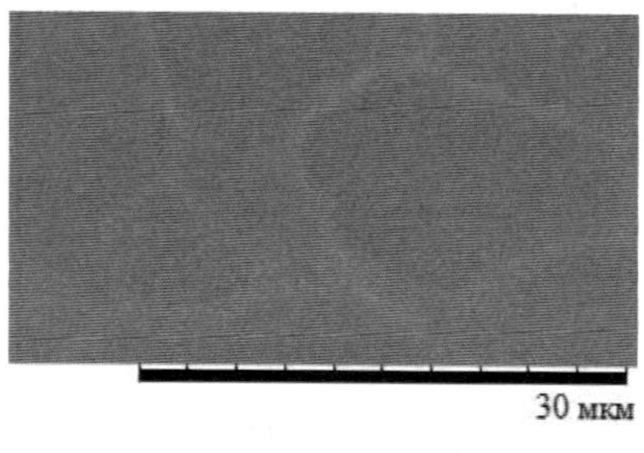

Figure 4. Film surface micrograph

For composition 3, the coating thickness was determined using the ellipsometry method. At a silicon concentration of 5 mg/ml, the film thickness is 78 nm, an increase in the content of silicon ions in the system to 20 mg/ml increases the coating thickness to 160 nm.

4. Conclusion

The main feature of floodplain lakes in Western Siberia in the summer and autumn is the presence of iron compounds and dissolved organic acids, which concentration in the study period did not change.

The determining factors of the water dynamics of iron are the concentration of humic acids, pH, CO_2. In summer, living activity is more intense, living organisms produce large amounts of CO_2, so the concentration is high. In autumn, the flora and fauna begin to die off and go into hibernation, but the photosynthetic speed is still high. Thus, regardless of the concentration of DOC, colloidal solutions are formed that are resistant to the settling process. For the formation of a stable colloidal system, a concentration of silicon ions of 10 mg/ml is sufficient. It was found that film formations are formed on the surface of natural waters, which have a thickness of up to 160 nm depending on the concentration of silicon ions in the system.

References

[1] Viers J, Barroux G, Pinelli M, Seyler P, Oliva P, Dupre B and Resende Boaventura G 2005 *Sci Total Environ* **339** 219
[2] Hutchinson G E 1969 A treatise on limnology: Geography, physics, and chemistry (Moscow: Progress Publ.) 591
[3] Alekin O A 1953 Basic concepts of hydrochemistry (Leningrad: Gidrometeorologicheskoye Publ.) 296
[4] Schelker J, Trauth N, Battin T J, Schmidt C 2019 *Hydrological Processes* **33** 2279
[5] Reiman J H, Xu Y J, He S, DelDuco E M 2018 *Chemosphere* **205** 559
[6] Chen C., Kukkadapu R K, Lazareva O and Sparks D L 2017 *Environ. Sci. Technol.* **51**(14) 7903
[7] Makkaveev N I 2003 Riverbed and its erosion in the basin (Moscow: Moscow State University Publ.) 355
[8] Mosimane K, Struyf E, Gondwe M J, Frings P, van Pelt D, Wolski P, Schoelynck J, Schaller J, Conley D J and Murray-Hudson M 2017 *Water SA* **43**(1) 104
[9] Shim M J, Cai Y, Guo L, Shiller A M 2017 *Hydrological Processes* **31**(5) 1086
[10] Karlsson T, Persson P 2012 *Chem. Geol.* **19** 322
[11] Sundman A,Karlsson T, Sjöberg S, Persson P 2014 *Geochim. Cosmochim. Acta* **145** 297
[12] Sundman A, Karlsson T, Persson P 2013 *Environ. Sci. Technol.* **47** 8557
[13] Noréan K, Loring J S, Bargar J R, Persson P 2009 *J. Phys. Chem.* **113** 7762
[14] Ioganzen B G, Tumentsev N F 1968 *The Ob floodplain: nature, reclamation, melioration* (Novosibirsk: Zapadno-sibirskoye knizhnoe izdatel'stvo Publ.) p 157 (In Russian)
[15] Electronic weather archive [Electronic resource]. URL: https://rp5.ru
[16] Petrov I B 1979 *The Ob-Irtysh floodplain. Typification and quality evaluation* (Novosibirsk: Nauka, Siberian Branch Publ.) p 136 In Russian

IOP Conf. Series: Materials Science and Engineering **696** (2019) 012003 doi:10.1088/1757-899X/696/1/012003

Transformation of Arctic ecosystems under impact of open-pit extraction of mineral resources

S B Selyanina, T I Ponomareva, G V Mikhailova, E Yu Churakova, I N Zubov and O N Yarygina

FCIArctic RAS, Arkhangelsk, Russia

E-mail: gumin@fciarctic.ru

Abstract. The article considers the influence of open-pit mining of diamonds on the state of Arctic ecosystems outside of the zone of the direct impact of the enterprise. To assess changes in natural complexes, a retrospective analysis was carried out on the basis of the data from social surveys of the local population using adapted methodological approaches and comparing descriptions of the biodiversity of aquatic macrophytes made in 1969-1970 and 2018. The comprehensive studies of peat soils and the laboratory modeling of the flow of the saponite into the watercourse were carried out, in order to understand the possible ways of the impact of diamond mining on natural complexes. The shallowing of the river, the change in the nature of soils in the lower reaches of the river, the crowding out a number of native species of aquatic plants by invasive macrophytes were educed. It is shown that saponite interacts with components of bog ecosystems, changes its physicochemical properties and forms intermolecular associates with dissolved organic matter. This contributes to the transport of mineral particles over long distances, to its accumulation in bottom sediments, and to the prolonged desorption, which creates favorable conditions for the introduction of invasive species.

1. Introduction

Mineral resources are a significant potential of the Arctic. Therefore, extraction of mineral resources is actively developed in the Arctic zone of the Russian Federation. In particular, the industrial development of diamond deposits by open-pit mining is conducted in the Arkhangelsk diamondiferous province. The mining and processing plant, which is developing the kimberlite tube named after M.V. Lomonosov, is located in the upper course of the Zimnaya Zolotitsa River. This river serves as a spawning site for salmon and other fish, as well as the main source of water for the local population [1]. The river flows along the whole length in the deeply carved valley, the soils in the upper course are rocky, and in the lower course mostly sandy. The river has mixed feeding from snow, rain, and groundwater [2]. PJSC "Severalmaz" carries out regular environmental monitoring in the impact zone – close to the enterprise [3].

The mass media outlined an extremely negative forecast for the transformation of these territories after the start of mining in 2003 [4,5]. It was not confirmed. At the same time, the consequences of diamond mining by an open-pit method for the lower reaches of the Zimnaya Zolotitsa River were practically not studied. It is of great interest to scientists and members of the public; it is important for predicting the possible impact of further development of mineral resources in the northern territories of the Russian Federation on natural systems and the quality of human life. The multiplicity of the factors of impact and the diversity of objects that are affected requires an integrated approach to assessing changes in the environment.

Therefore, the purpose of this study was to identify changes in the state of natural complexes as a result of open-pit mining of mineral resources (the case of the catchment area of the Zimnaya Zolotitsa River outside the official zone of impact of the diamond mining enterprise) using sociological and natural scientific methods.

2. Methods and materials

The studies were carried out in in the catchment area of the Zimnaya Zolotitsa River. The study area is situated in the north of the mainland part of the Arkhangelsk region, within the Belomorsko-Kuloisky plateau, in the zone of tundra forests. There is a quarry mining of a diamond deposit in the upper course of the Zimnaya Zolotitsa River. The main factors of the mining and processing plant direct impact on the environment are a decrease in the level of groundwater and a change in the hydrological regime, aeolian transfer of solid particles from terricones, water entering the riverbed from sedimentation ponds and quarry. Discharged waters containing finely dispersed clay mineral – saponite, according to the technological cycle of the enterprise, are treated on the filtration fields – plots of the high-moor sphagnum bogs, which occupy large areas in this territory [6].

Field materials were collected in 2018 in the lower course of the Zimnaya Zolotitsa River (at a distance of more than 100 km downstream) and in 2019 in its upper course (close to the plant).

A retrospective analysis of the changes that occurred during the industrial development of the quarry was made based on data from social surveys of the local population using questionnaire survey methods, expert and group interviews (focus groups) [7].

The main object of floristic studies was true water plants, which can serve as a reliable indicator of changes in the state of water bodies [8, 9]. Analysis of the transformation of plant communities in the region out based on data from 1969-1971 [10], and from field materials received in 2018-2019 was carried.

To understand the possible ways of mining developments affects natural complexes, field studies of peat deposits in the bogs of the catchment area and the laboratory modeling of entering the watercourse of the saponite suspension, both directly and through filtration fields (oligotrophic peat bogs), were carried out. Resistance to sedimentation was evaluated by changing the optical density of model waters of various concentrations, measured using a UV-spectrophotometer [11]. The changes in physicochemical parameters of peat deposits: active acidity, redox potential (ORP), total mineralization and degree of decomposition were studied [12].

3. Results and discussion

Results of surveys of the residents showed that most of them are not satisfied with the state of the environment. They noted that the water in the reservoirs is contaminated and the quality of drinking water is poor. At the same time, half of the residents use this water for food purposes due to the lack of other sources of water. Most respondents consider the mining and processing plant activity as the cause of water pollution. People in the survey noted overgrowing of the riverbank with shrubs and trees, erosion of coastal slopes, and a change in the composition of the vegetation cover. They associate these changes with a reduction in haying, grazing, etc. Figure 1 shows the results of surveys of the residents.

Participants in interviews and focus-groups noted shallowing of the river and lowering of groundwater levels; changes in the nature of soils in the lower course of the river; changing the terms of freezing and ice drift. Three cases, over the past seven years, have been noticed when water assumed a pronounced red-pink color. That is characteristic of kimberlite ore from the Arkhangelsk diamondiferous province containing saponite. It is significant that the villages of Verkhnaya and Nizhnaya Zolotitsa are 100-120 km away from the mining and processing plant. According to the population opinion, due to the deterioration in the water condition, salmon rarely spawn in the river, and the number of individual species of river fish decreases.

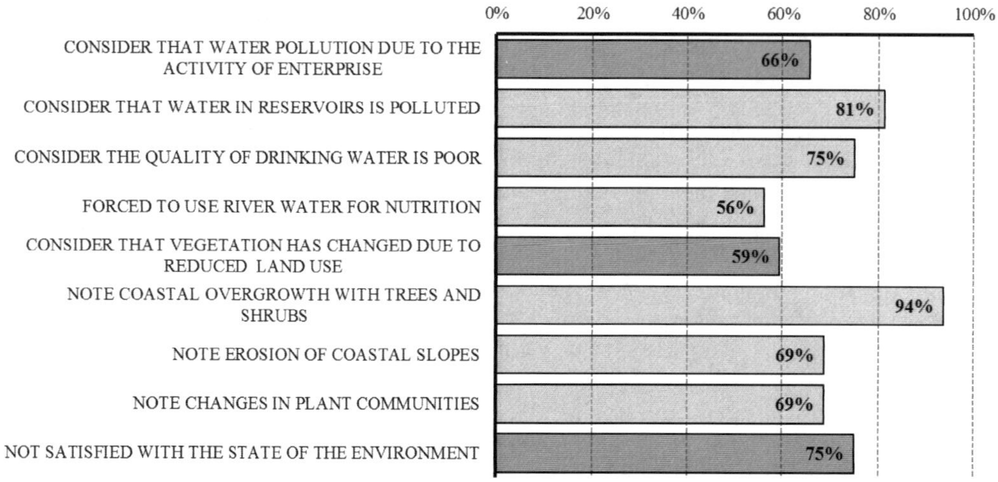

Figure 1. The results of a survey of the population of the Verkhnaya Zolotitsa village.

In witness whereof, there were recorded changes in the composition of aquatic macrophytes during the assessment of plant societies of studied territories. Their diversity, in comparison with the observations of 1969-1970, increased to 17 species in 2018. The specimens of *Myriophyllum spicatum L.* and *Persicaria amphibia (L.) Delarbre* were not found in the riverbed, but earlier these species were widespread. Four species of true aquatic vascular plants were noted for the first time: *Elodea canadensis Michx., Potamogeton lucens L., Stuckenia filiformis (Persoon) Börner* and *S. pectinata (L.) Börner*. The first of them forms dense thickets in some areas and is able to displace native species of aquatic macrophytes. For the European Plain, findings of these species in the surveyed area are located further to the north than all previously mentioned [13]. Most species whose frequency of occurrence has decreased are characteristic of weakly mineralized water bodies. Newly recorded species, on the contrary, prefer waters with increased salinity and hardness [14] and are demanding for nitrogen availability [15,16].

The active dispersal of new types of hydrophytes is probably facilitated by deoxidation of waters and an increase in their salinity. This can be caused by a strong decrease in water level in general, especially in the summer low water season, both due to natural climatic changes [17], due to a change in the species composition of forests as a result of fires and logging in the middle course of the river, as well as under the impact of mining.

Physico-chemical characteristics of peat deposits at different distances from the mining and processing plant are shown in Figure 2.

It should be noted that on the first and second plots between the tirr and the body of the peat deposit there is a thin (2-5 cm) highly oxidized black layer of weakly decomposed plant residues, and the peat itself has significantly higher density in comparison to the areas remote from the enterprise. Moreover, in the immediate vicinity of the enterprise, the consequences of the aeolian transfer of saponite are traced. The true mosses dominate in the moss layer of the plots which are situated near the terricones, and sphagnum mosses – in the distance. On the first plot (Figure 2 a) the deoxidizing effect of saponite introduction in the upper layers of peat (up to 15-25 cm) is manifested. On the other plots, the upper layer is characterized by the higher acidity (pH = 3.2-3.6), which decreases slightly with the depth of the deposit (Figure 2b, 2c).

The presence of saponite promotes the oxidation of peat components, which demonstrates the redox potential characteristic curves (Figure 2 d-f). At the same time, the barrier on the way of aeolian transport in the form of a tree belt area of 150-200 m width significantly reduces the impact of saponite on the bog massif.

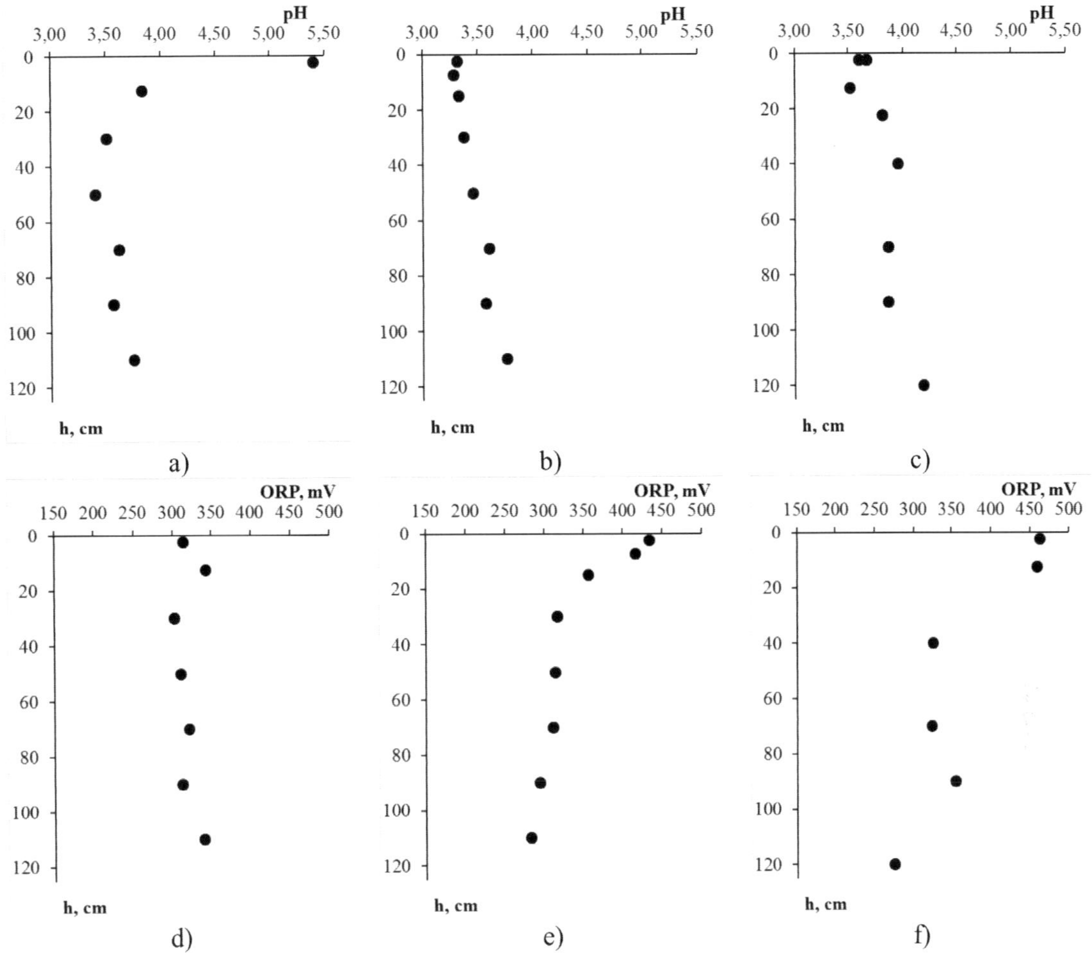

Figure 2. Physico-chemical characteristics of peat deposits: a), d) - in the immediate vicinity of the enterprise; b), e) - at a distance of 1 km from the enterprise on a site shielded by forest; c), f) - at a distance of 50 km from the enterprise.

According to laboratory studies, the deoxidizing effect of saponite is proportional to its addition to bog water. A decrease in active acidity (increase in pH value from 3.2-3.6 to 4.1-4.3) occurs already at short observation times. When filtering through peat, clarification of the suspension occurs due to the mechanical retention of saponite particles by parts of sphagnum mosses. During the first day, the peat under study provided a decrease in the total mineralization of water by 40-64 %, on the second day the degree of purification was 11-16 %, and on the third day it stabilized at a level of 2-6 %. By the end of the 5th day of the model experiment, a slight increase in total mineralization was observed. This indicates the exhaustion of the sorption capacity of peat. The optical density of bog water with saponite additives less than 0.01 % and subsequent sedimentation (5 days) increases by 1.2-1.5 times. An increase in saponite supplementation causes a proportional increase in the indicator. In this case, Springer's chromaticity coefficient (the ratio of optical densities at wavelengths $\lambda = 465$ nm and $\lambda = 665$ nm) increases slightly in the presence of saponite but does not depend on the weight of the additive. Saponite is characterized by a maximum optical density in the wavelength range $\lambda = 420$-470 nm [18]. The revealed effect is apparently due to the formation of sedimentation-resistant intermolecular organo-mineral associates of a certain composition. Such particles can be transported by the water stream at considerable distances from the source of pollution.

4. Conclusion

The result of the studies is that changes were recorded in the state of the natural complexes of the Zimnaya Zolotitsa River – outside the zone of the direct impact of open-pit mining. That is, first of all, a change in the level and composition of ground and river waters, changes in bog, land and aquatic plant communities, as well as in the physicochemical properties of peat deposits.

The change of plant communities occurs both under the influence of mining (aeolian and water transport), and other (climatic and anthropogenic) factors.

Peat has a high buffering ability with respect to saponite, but it significantly changes its physicochemical properties. Moreover, with a high degree of probability, it can be spoken of the formation of intermolecular associates of the mineral with humic substances, which contributes to the transfer of saponite particles over long distances, their accumulation in bottom sediments, as well as the prolonged desorption of mineral substances. This, in turn, creates favorable conditions for the introduction of invasive species.

The data obtained are important both for understanding the ongoing changes and for planning remediation activities.

Acknowledgments

The reported study was funded by RFBR according to the research project № 18-05-60151

References

[1] Aksenov D E, Dobrynin D V, Dubinin M YU, Egorov A V, Isaev A.S., et al. 2003. *Atlas of Russia's Intact Forest Landscapes* (Moskow: MSoES) [in Russian]
[2] Artemov V V 2010. *Technical report RosGidromet.* (Arhangelsk: GU «Arhangelskij CGMS-R») [in Russian]
[3] Annual report of PJSC «Severalmaz» in 2017. Accessed at URL: http://www.severalmaz.ru/investoram/godovye-otchety/[in Russian]
[4] Ecology: Arkhangelsk region will pay for diamonds by the White Sea. Ecology of production. Scientific and practical journal. 2006. Accessed at URL: http://www.ecoindustry.ru/news/view/8468.html [in Russian]
[5] The public demands environmental transparency of the Lomonosovskoye deposit from Severalmaz. Information and analytical portal ROSNORD. 2007. Accessed at URL: http://rusnord.ru/oldnews/10151-24348.html [in Russian]
[6] Sheveleva A V, Shvartsman Yu G 2012 *Arctic Evironmental Research.* **2** 40 [in Russian]
[7] Eparhina O V 2013. *Sociology of public opinion* (Moscow: Akademiya) [in Russian]
[8] Papchenkov V G 2003 *Hydrobotany: methodology, methods: materials of the hydrobotanic School (Borok, 8–12 April 2003)* 23 [in Russian]
[9] Poikane S, Portielje R, Denys L, Elferts D, Kelly M, et al. 2018 *Ecological Indicators.* **94** 185 DOI:10.1016/j.ecolind. 2018.06.056
[10] Dobrovol'skaya T G 1971 *Materials for the analysis of the flora of the vicinity of the village Zimnyaya Zolotica of the Primorsky district Arkhangelsk region* (graduation thesis, head A. I. Tolmachev) (Leningrad) [in Russian]
[11] Khodakov G S, Yudkin Yu P 1980 *Sedimentation analysis of highly dispersed systems.* (Moscow: Nauka) [in Russian]
[12] Lishtvan I I 1989 *Physics and Chemistry of Peat: Proc. manual for high schools* (Moskow: Nedra) [in Russian]
[13] Bobrov A A, Chemeris E V 2009 *Reports of the Academy of Sciences.* **425(5)** 705 [in Russian]
[14] Bobrov A A, Cel'movich O L, Otyukova N G 2006 *Hydrobotany 2005 (Borok, 11–16 October 2005)* 210 [in Russian]
[15] Landolt E 1977 *Veroff. Geobot. Inst. ETH.* **64** 1

[16] Zhukova L A, Dorogova Y A, Turmuhametova N V, Gavrilova M N, Poljanskaja T A 2010 *Ecological indicator values and methods of analysis of ecological diversity of plants*. (Mari State Univ. Yoshkar-Ola) [in Russian]

[17] Zelenina L I , Nesmelova A I 2015. *Modern scientific research and innovation*. **3(2)**. Accessed at URL: http://web.snauka.ru/issues/2015/03/50196.

[18] Kanygina O N, Kravtsova O S, et al. 2011 *OGU Bulletin* **12(131)** 393 [in Russian]

IOP Conf. Series: Materials Science and Engineering **696** (2019) 012004 doi:10.1088/1757-899X/696/1/012004

Comparative assessment of drag reduction efficiencies of polymer solutions and surfactants at low temperatures

E S Kashlach[1], E M Berezina[1], A S Smirnova[1], I A Berezina[1], V N Manzhai[1,2] and M S Fufaeva[2]
1National Research Tomsk State University, 634050 Tomsk, Lenin Avenue, 36
2Institute of Petroleum Chemistry, Siberian Branch of the Russian Academy of Sciences (IPC SB RAS), 4, Akademichesky Avenue, 634055, Tomsk, Russia

E-mail: maria81@ipc.tsc.ru

Abstract. The antiturbulent efficiency of solutions of high molecular weight polymers and that of micellar systems of surfactants are compared. The effect of the structural organization of polymer solutions and colloidal systems on their antiturbulent efficiency is shown. Hydrodynamic and physicochemical approaches are used to interpret the experimental results. It is shown that the use of polymers as drag reducing agents in main oil pipelines is more preferable at low liquid temperatures in northern latitudes (the Arctic).

1. Introduction

By dissolving the polymers, homogeneous solutions are formed, which have some properties of colloid systems (low diffusion rate of polymer macromolecules with immobilized solvent and dispersed phase particles of colloid systems, small values of osmotic pressure, etc.). This can be explained by similarity of their structural-mechanical characteristics and the large size of polymer molecules and micelles ($\sim 100 \cdot 10^{-9}$ m), in comparison with to the small molecular size ($\sim 1 \cdot 10^{-9}$ m) of the dispersion medium (solvent).

One more feature that binds colloid particles with macromolecules is their ability to reduce hydrodynamic resistance. This phenomenon, called the Toms effect (TE), manifests in increasing turbulent flow rate of the polymer solutions in comparison with the flow rate of the initial solvent [1]. It has also been found that colloid systems, having aqueous or hydrocarbon dispersion medium, are also capable to reduce hydrodynamic resistance.

The Toms effect is currently widely used in various fields, for example, in the oil industry. The oil-soluble polymers of different chemical nature (polyisoprene, polybutadiene, polyisobutylene, polyolefins, etc.) are used to intensify the industrial pumping of oil and petroleum products. The turbulent flow solutions of hydrocarbon liquids regularities are of practical and theoretical interest as well.

The aim of this paper is to establish functional connections between hydrodynamic flow parameters, physical and chemical properties of the solutions, their structure and antiturbulent efficiency of polymer solutions and surface-active substances (SAS). Transferring the TE on the quantitative level will allow us to estimate *a priori* the efficiency of antiturbulent additives.

2. Experimental section

Laminar flow of liquid in the cylindrical channel becomes turbulent at a Reynolds Number that is greater than the critical value $Re_* = 2UR_W\nu^{-1} \geq 2300$. Changing a flow regime is accompanied by a coefficient increase in hydrodynamic resistance **λ**. This leads to the energy expenditures growth for pumping liquids in the turbulent regime. Regardless of the flow regime, the functional linkage between the volume flow rate (*Q*) and set pressure drop (*ΔP*) is described by the Darcy-Weisbach equation

$$\Delta P = \lambda \frac{L}{4\pi^2 R_W^5}\rho Q^2, \qquad (1)$$

where $Q = \pi R^2 U$ – volume flow rate; *U* – the average linear liquid velocity; $\Delta P = P_1 - P_2$ – friction pressure drop in the pipe with the length *L* and radius R_W; *ρ* – the liquid density; *η* and $\nu = \eta / \rho$ – dynamic and kinematic viscosity, respectively.

The fluidity experiment testing of various liquids (one-component solvents, polymer solutions and colloid systems) has been carried out using a laboratory turbulent rheometer (tube radius $R_W = 1.1 \cdot 10^{-3}$ m, length L = 0.8 m). A turbulent rheometer is similar in design to a capillary viscometer and that allows us to take measurements in laminar and turbulent flow regime [2]. By passing the liquid through a cylindrical channel during the experiment, the turbulent rheometer measured volume flow rate *Q* and pressure drop *ΔP*. The viscosity values, density, and temperature of solutions and solvents are known prior to the experiments. The polyisoprene solutions (a repeating unit formula of the polymer chain ($–CH_2–CH=C(CH_3)–CH_2–)_n$) in petrol (concentration C = 100g/m^3 and C = 200g/m^3, a molecular mass of $M_r = 1{,}1 \cdot 10^6$) and solutions of oil-soluble SAS (C = 1000 g/m^3, and C = 2000 g/m^3) have been investigated. As SAS, aluminum salts of higher carboxylic acids (formula $(AlOH(RCOO)_2)_n$, where R = $(CH_2)_{10-20}$) have been used. These SAS are also called «aluminum soap» (Al-soap). The results of hydrodynamic experiments (Figure 1) are represented as a variation $\lambda = f(Re)$.

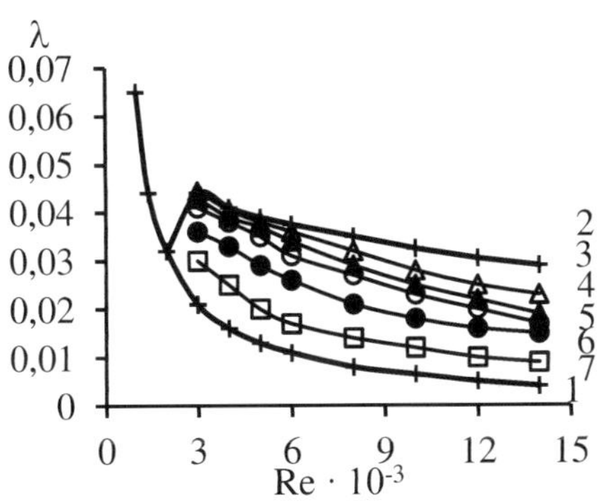

Figure 1. Variation of drag (hydrodynamic resistance) coefficient with Reynolds number for different systems: 1 – Poiseuille theoretical curve for the laminar $\lambda_{LAM} = 64 / Re$ $(Re \leq 2300)$ and turbulent flow $(Re \leq 2300)$; 2 – Blasius empirical curve for turbulent flow λ_{TUR} = $0.3164 / Re^{0.25}$; 3 – polyisoprene solution in gasoline (C = 100 g/m^3; M_r = $1.1 \cdot 10^6$); 4 – SAS solution (Al-soap) in gasoline (C = 1000 g/m^3); 5 – polyisoprene solution in gasoline (C = 200 g/m^3; M_r = $1.1 \cdot 10^6$); 6 – SAS solution (Al-soap) in gasoline (C = 2000 g/m^3); 7 – polyisoprene solution in gasoline (C = 200 g/m^3; M_r = $2.1 \cdot 10^6$).

The analysis of experimental data and variation $\lambda = f(Re)$ showed that the coefficients of hydrodynamic resistance *λ* in laminar flow region for Newtonian liquids (water, oil, gasoline, etc.) lie on the Poiseuille curve (Figure 1, curve 1). The transition to turbulent flow regime occurs in Reynolds number range from $Re \cdot 10^{-3} = 2{,}0 - 3{,}0$. It is accompanied by sensible increase in hydrodynamic resistance and the transition magnitude values *λ* from the Poiseuille curve onto the empirical Blasius curve (Figure 1, curve 2), typical for the flow of Newtonian liquids at $Re \cdot 10^{-3} > 0{,}3$.

Figure 1 shows that the current non-Newtonian liquids (polyisoprene solutions and Al-soap) in a turbulent regime reduce the drag coefficient *λ* (the Toms effect).

For laminar flow regime, values λ for the investigated solutions (as for Newtonian liquids) lie on the Poiseuille curve (curve 1), but for turbulent flow regime values λ are located in the area (curves 3-7), bounded by Blasius curve above, and the hypothetical extension of the Poiseuille curve in turbulent flow below. Thus, the drag reduction coefficient λ is acceptable only for the turbulent regime. The greater the concentration (C) of a dissolved additive and higher a molecular weight (M_r) of the polymer sample is, the greater a reducing effect λ is.

It should be noticed, that minimum values of drag coefficient λ of polymer solutions turbulent flow, are limited by the Virk asymptote $\lambda = 2,36 \cdot Re^{-0,58}$. It is shown, that under no circumstances they may fall within the area below the Poiseuille curve in the turbulent region (Figure 1, curve 1). In other words, the effect of antiturbulent additives comes up to a partial flow laminarization.

The Toms effect ($DR,\%$) is accepted to estimate quantitatively by relative change in the drag coefficient, expressed as a percentage:

$$DR,\% = \frac{\lambda_S - \lambda_P}{\lambda_S} 100\%, \tag{2}$$

where λ_S – the drag coefficient of a the pure solvent; λ_P – the drag coefficient of the solution with the additive.

The coefficient λ_S and λ_P can be calculated from the transformed Darcy-Weisbach formula

$$\lambda = \frac{4\pi^2 R_W^5}{L\rho} \cdot \frac{\Delta P}{Q^2}.$$

In accordance with the Darcy-Weisbach equation (1) a reduction in the drag coefficient λ provided $\Delta P = const$ should be accompanied by an increase in the volume flow of solution additives Q_P as compared to a rate of liquid without additives Q_S on the value $\Delta Q = Q_P - Q_S$. Using these experimentally determined values one can calculate the Toms effect ($\Delta P = const$) in the expression (3):

$$DR,\% = \left[1 - \frac{Q_S^2}{(Q_S + \Delta Q)^2}\right] 100\%. \tag{3}$$

The effect size of drag reduction with the Reynolds number increase approaches to its maximum value DR_{MAX} ~ 75%. The antiturbulent additives Baker, Necadd, X-Pand, Liquid Power on the base of oil-soluble high molecular weight polymers ($M_r \sim 10 \cdot 10^6$) (which are currently widely used in the world of industrial oil pumping) have such a high effect size. It is enough to add 5-10 grams (5-10 ppm) of polymer per 1 tonne of oil to achieve the maxi effect size.

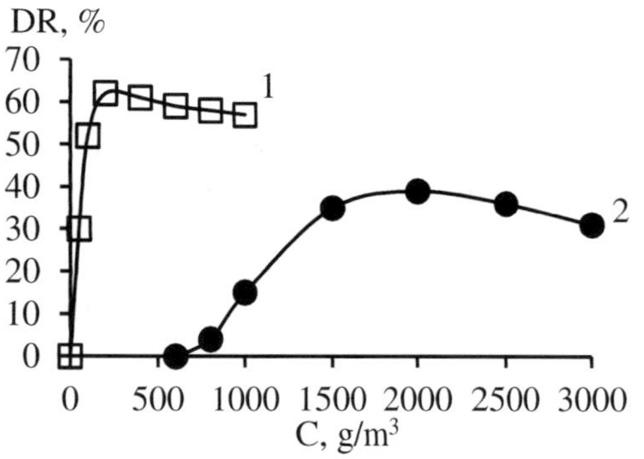

Figure 2. Variation of the Toms effect magnitude and antiturbulent additives concentration (polyisoprene solutions (1) and SAS (2) in gasoline).

To compare the efficiency of different antiturbulent additives, one can use a graph variation of the effect size on the additives concentration $DR = f(C)$. As it seen from a typical figure. 2, the effect size increases rapidly and reaches its maxi value (DR_{MAX}) at an «optimal» concentration (C_{opt}) when the additive concentration increase occurs. By further concentration increasing, the effect is being reduced. Figure 2 shows that it is required 10 times less polyisoprene ($C_{opt\text{-}1}$ = 200 g/m³), than the oil-soluble SAS ($C_{opt\text{-}2}$ = 2000 g/m³) for the maxi effect size. The smaller the value of «optimal» concentration is, the more effective antiturbulent additives are.

For this reason, antiturbulent additives based on polymers, versus SAS, find wide application in pipeline transportation of crude oil.

A significant drawback of the formulas (2) and (3) is the absence of strongly expressed functional dependence of the effect (*DR*) on the physicochemical properties of the solvent and solutions thereof. Moreover, these expressions do not allow us to forecast the effect size *a priori.*

3. Physicochemical description of the turbulent flow of polymer solutions and colloidal systems

To explain the reason of the different antiturbulent efficiency for polymers and SAS, we should drill down the structure of polymer solutions and colloid systems. The macromolecules, which consist of a large number of monomeric units, linked by covalent bonds, have a specific structure in a solution. The polymer chains are coiled in loose skeins, which volume depends on molar mass of the polymer (*M*) and intrinsic viscosity $[\eta]$ of polymer solution. It can be calculated by the formula $V_k = M[\eta]/N_A$, which is a consequence of the well-known equation of Fox-Flory $(h^2)^{3/2} = M[\eta]/\Phi$.

The coils of macromolecules always hold within themselves a certain amount of a solvent. Each monomeric unit of a polymer chain, depending on the thermodynamic quality of the solvent, is surrounded by dozens or even hundreds of the solvent molecules. According to the geometric dimensions, the solvent molecules are comparable to monomeric units. In the volume of the macromolecular coil with immobilized solvent at a fraction of the polymer chain accounts not more 1% of the total mass. Therefore, the density of swollen coil (ρ_k) and solvent (ρ_s) are approximately equal ($\rho_k \approx \rho_s$). Consequently, a macromolecular coil can be regarded as the solvent microvolume which is structured by a polymer chain. Such a microstructure has not only viscous, but also elastic properties. In other words, an individual macromolecule with immobilized solvent can be regarded as a disperse system «microphase». The elasticity of macromolecular coils, unlike micelles elasticity, has entropic nature and can be calculated by the known formula in polymer rheology $G \approx kT/V_k$. Therefore, the coefficient of elasticity (*G*) of macromolecular coil with the immobilized solvent, increases with the temperature (*T*) and decreases with increasing coil volume (V_k).

An equation is suggested, relating the increment in the volumetric velocity gain (ΔQ) with hydrodynamic flow characteristics and physicochemical properties of the polymer solution in the work [3, 4]:

$$\Delta Q = \pi R_W^2 \Psi \frac{\tau_W}{\sqrt{\rho G}} \left(1 - \sqrt{\frac{G}{\tau_W}}\right), \qquad (4)$$

where $\Delta Q = Q_P - Q_S$ – volume flow increment in the polymer solution (Q_P) as compared to the volume rate of the pure solvent (Q_S) at equal shear stresses on the tube wall ($\tau_P = \tau_S = \tau_W$); $\Psi = C[\eta]/(1 + C[\eta])$ – volume fraction of macromolecular coils, depended on the concentration of dissolved polymer (*C*); $\tau_W = (\Delta P/2L)R_W$ shear stress on the tube wall; ρ – solvent density; $G = kT/V_k$ – coils' elasticity coefficient; *k* – the Boltzmann constant; V_k – the macromolecular coils' volume; T – the absolute temperature.

The equation is derived theoretically based on a hypothetical model, which is based on the assumption, that the layered (shear) liquid flow is caused by rolling of the particles (molecules or associates) of one layer by the adjacent layer surface, but lying close to the wall of the pipe. The particle takes part in rotational and translational motion in which it is deformed periodically by flowing around its layers (following and overtaking) in the process of rolling. When developed turbulent flow with high Reynolds numbers and, consequently, high values of shear stresses ($\tau_W >> G$) the equation (4) takes the form

$$\Delta Q = \pi R_W^2 \Psi \frac{\tau_W}{\sqrt{\rho G}}, \qquad (5)$$

The coils elasticity coefficient *G* in the equation (5) is crucial, which characterizes the elastic properties of macromolecular coils with immobilized solvent. Therefore, the smaller the elasticity of macromolecular coils is, the greater the volumetric speed gain of polymer solution turbulent flow is, in comparison with the flow rate of the pure solvent. In its turn, the lower the coils elasticity is, the more their volumes (molecular mass) are. The experimental data on the dependence of volumetric speed gain (ΔQ) on various parameters (τ_W, T, C), which confirm the validity of the theoretical equation (5) are shown in paper.[5]

The volume flow increment ΔQ is included in the equation (3), so the increase in this value must be accompanied by an increase in the effect of drag reduction *DR*. The function of $\Delta Q = f(C)$,as well as the dependence of $DR = f(C)$ for dilute solutions, with concentration growth, is increased according to linear law, reaching its maximum at the same value of «optimal» concentration (C_{opt}) and then slowly decreases in the concentrated solutions areas.[5]

It is known that the properties of polymer solutions are determined not by the value of the concentration of dissolved polymer (C), but the value of the volume fraction $\psi = \sum V_k / V_O$ occupied by all the macromolecular coils in the total volume of the V_O. The structure ofthe polymer solution changes depending on the volume fraction of macromolecular coils if the dissolved polymer concentration increases.

Figure 3 schematically demonstrates polymer solutions of different concentrations. The macromolecules are separated by solvent layers and have unperturbed equilibrium dimensions ($V_k = const$) in the dilute solution with volume fraction of coils smaller than one (Figure 3, state 1). «Maximally dense packing principle» is realized with increasing concentration of macromolecules in a certain moment (Figure 3, condition 2), where in the volume fraction of the coils in the solution becomes approximately equal to one $\psi \approx 1$. The macromolecular coils with immobilized solvent are in close contact, but they do not deform each other yet. This concentration is called «optimal». With further concentration increase the coils begin to discriminate against each other, i.e. become compressed and reduced in volume. In the result we observe the polymer component content increase in the coil and its rigidityincreases as well. Thus, the changes of the solution structure should lead to a change in its fluidity that is predicted by the equation (5) and confirmed experimentally.

In a dilute solution ($\psi << 1$) due to the low concentrations (C) value of $C \cdot [\eta] << 1$, and then the denominator of the formula $\Psi = C \cdot [\eta] / (1 + C \cdot [\eta])$ in the equation (5) is equal to one. Consequently, a velocity increment's linear dependence on concentration should be performed for the dilute solution

$$\Delta Q = \frac{\pi R_W^2 \tau_W}{\sqrt{\rho G}} [\eta] C = const \cdot C.$$

There is a compression of coils and increase of their resilience in the concentrated polymer solution (Figure 3, state 3) for $C \cdot [\eta] >> 1$ and $\psi = 1$. This should be accompanied by a decrease in the volume flow increment proportionally $G^{1/2}$, i.e. the monotonous drop effect takes place in accordance with $\Delta Q = \frac{\pi R_W^2 \tau_W}{\sqrt{\rho G}} = \frac{const}{\sqrt{G}}$.

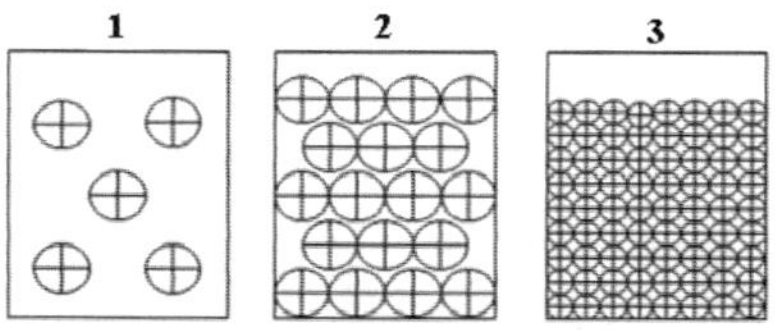

Figure 3. The structure of the polymer solution: 1 – a dilute solution ($0 < \Psi < 1$; $C < C_{opt}$, $V_k = const$); 2 – an «optimal» concentration solution ($\Psi \approx 1$; $C = C_{opt}$, $G_1 V_k = const$; $G_2 \approx G_1$); 3 – a concentrated solution ($\Psi = 1$; $C > C_{opt}$, $V_3 < V_k$; $G_3 > G_1 \approx G_2$)

These theoretical considerations are confirmed experimentally for all known twains a polymer – a solvent. The «optimal» concentration C_{opt} differs for various polymers. The greater the volume of coils is, the less C_{opt} is, and it is more profit able to use high molecular weight polymer sample in industrial practice.

The main drawback of high-molecular polymers ($M_r > 1 \cdot 10^6$) is their irreversible destruction when they are injected into the flow with high shear stress if they are used as drag reducing agents. Breaking covalent links between adjacent monomeric units is accompanied by a decrease in polymer chain length and coil volume, which reduces the Toms effect until its complete disappearance [5]. The polymer destruction is signified during liquids passage through industrial pumps on main product pipelines. So, it is necessary to add a fresh portion of antiturbulent additives after the passage of liquids through the pumping stations.

A somewhat different picture is observed in colloid systems. Surface-active substances dispersed (dissolved) up to separate molecules of relatively small molecular weight ($M_r \approx 200 - 500$) in dilute solutions. By increasing the content of SAS and after reaching the critical concentration for micelle formation (CCM) in the solution, the aggregates are formed. The aggregates composed of a large number of SAS molecules, which are structured in ordered associates (micelles). Consequently, achieving CCM, the homogeneous SAS solutions drop into heterogeneous systems. It was established experimentally that the hydrodynamic resistance may reduce SAS solutions only after reaching the CCM, i.e. the concentrated solutions, where large enough micelles have already been formed.

Surface-active substances molecules possess a polar group and a non-polar hydrocarbon radical, i.e. have a diphilic structure. Therefore, the micelle structure depends on the polarity of dispersing medium. For example, in a hydrocarbon (non-polar) dispersing medium the polar functional groups of the SAS molecules are located in the micelle core and linked by intermolecular forces (the ordering and strength of arising associates depend on them). The hydrocarbon «tails» of diphilic SAS molecules on the micellesperiphery provide solubility in the nonpolar medium. The strength of such supramolecular structures is low because of the low intermolecular interaction energy. But due to the tight convergence of SAS molecules in the formed micelles, their coefficient of elasticity G_{SAS} is much greater than the polymeric macromolecular coils coefficient of elasticity ($G_{SAS} >> G_{POLY}$). Therefore, in accordance with equation (5), the effectiveness of antiturbulent rigid micelles is significantly lower than with the macromolecular coils with immobilized solvent. There is no condition 3 in micellar solutions. Instead, the volumetric structured system is formed that can run only after its destruction under large external force.

4. Conclusions

1. On the basis of the thermodynamic and physicochemical approaches, the functional dependence of polymer solutions' and surface active substances' antiturbulent efficiency on the hydrodynamic flow parameters, physicochemical solutions properties and their structures is established.
2. The reasons of varying antiturbulent efficiency of polymers and surface active substances were validated theoretically. The high antiturbulent efficiency of polymer solutions is the result of a low elasticity of macromolecular coils which elasticity has the entropic nature. Significantly less efficiency of colloid systems is due to the high rigidity of the micelles of surface-active substances solutions which elasticity has the energy nature.

Acknowledgments
The work was performed within the framework of project V.46.2.3. of the Program for Basic Scientific Research of State Academies of Sciences.

References

[1] Toms B A 1949 *Proceeding International Congress on Rheology* 2 p 135
[2] Malkin A Ya, Nesyn G V, Ilyushnikov A V, Manzhai V N 2001 *J. Non-Newtonian Fluid Mech* 97, p 195
[3] Manzhai V N, Nasibulina Yu R, Kuchevskaya A S, Filimoshkin A G 2014 *Chemical Engineering and Processing: Process Intensification* 80, p 38
[4] Manzhai V N, Sarycheva G A, Berezina E M 2003 *Polymer science series B* 45, p 45
[5] Manzhai V N, Nasibulina Yu R, Abdusalyamov A V 2015 *Russian Journal of Applied Chemistry* 88, p 118

Structure and properties of invar alloys for MIM technologies

S N Kulkov[1,2,3], E V Abdulmenova[2,3], V S Shadrin[1], A A Leonov[3,4] and R V Levkov[3]

[1] National Research Tomsk State University, Tomsk, Russia
[2] Institute of Strength Physics and Materials Science SB RAS, Tomsk, Russia
[3] National Research Tomsk Polytechnic University, Tomsk, Russia
[4] Institute of High Current Electronics SB RAS, Tomsk, Russia

E-mail: Ekaterina.V.Abdulmenova@yandex.ru

Abstract. The paper studied Fe-Ni alloys obtained by sintering from mixtures of powders. Powders with 30-40 wt. Ni were mixed with paraffin and wax binder and then pressed and sintered at 1300 °C during 1 hour. It has been shown that the face centered cubic lattice (γ-phase) and body centered cubic lattice (α-phase) are formed after sintering of powders mixture. The content of the α-phase decreases from 37 % to 2 % with increase of Ni concentration from 30 to 35 wt. % Ni in powders mixture. The average grain size of the γ-phase decreases from 26.5±12.3 μm to 18.3±9.1 μm, and porosity of samples increases from 13±1.4 % to 18.4±1.1 %.The lattice parameter of the γ-phase increases from $3.5849 \cdot 10^{-10}$ m to $3.597 \cdot 10^{-10}$ m and it corresponds to well-known invar alloys. The maximum value of the microhardness of sintered samples is 1.04±0.07 GPa. Thus, alloys obtained by sintering of Fe and Ni powder mixtures has an invar properties typical of cast alloys.

1. Introduction

One of the most important strategic trend for the development of Siberia and the Arctic is the creation of new materials. Accordingly, there is a need to develop new materials and technologies for their manufacture with specific mechanical or functional properties. Fe-Ni alloys with low coefficient of thermal expansion (CTE) [1, 2] has demand for the application in the field of high precision devices such as optoelectronic or light wave communication systems, housing or structural components [3]. There are different methods for obtaining of invar alloy (iron-nickel alloy), but powder metallurgy (PM) presents several processing advantages as compared with casting method.

Metal injection molding (MIM) is one of the main effective methods for fabrication of product with complex geometry [4]. Usually, it uses Fe-Ni alloy particles obtained by evaporation and cryo-condensation process and the gas-phase method [5-7] for realization of this technology, but these technique is rather expensive for obtaining of alloy powders. Powder metallurgy allows obtaining alloys with different composition by sintering from powders mixtures; however, there are not enough studies of invar alloys sintered from powders mixtures. Nevertheless, these studies are important for 3D-printing and MIM technologies.

The aim of this paper is to study structure and properties of alloys obtained by sintering from powders mixtures which can be used for MIM technologies.

2. Material and methods

The Fe-Ni mixtures of iron and nickel powders were used, the nickel content varied from 30 to 40 wt. %. The preparation of samples was carried out on several steps [8]. First of all, it was prepared binder that consists of paraffin and wax in a ratio of 90:10 which was mixed for 10 min. After this, the metal powder was added to binder, mixed for 30 min and this mixtures were uniaxial pressed. The samples were sintered in a vacuum furnace; the sintering temperature was chosen according to the phase diagram of the Fe-Ni alloy and was 1325 °C in 1 hour, heating rate was 5 °C min^{-1}, before sintering process samples were hold at 200 °C where binder was removed during one hour of holding time. The samples were etched with a solution of one part nitric acid in 20 parts ethanol [9].

The average grain size of the γ-phase and porosity of the finished MIM-specimens was studied by light microscopy (Labomet-I, Russia). The phase identification was investigated by the X-ray diffraction (XRD). XRD measurements were carried out using a difractometer with Co Kα radiations (Kα = 1.7902 Å). X-ray patterns were obtained using step-by-step method with statistics accuracy better than 0.5 %. XRD line profile analysis was used to determine the coherently diffracting domain (CDD was calculated by the Scherrer equation [10] for the lines (111)) and microdistortion (calculated according to the Stokes – Wilson equation [11] for the lines (311)). The lattice parameters were calculated with extrapolation on $\cos^2\theta$ values [12] using all lines. Finally, the microhardness of the samples was obtained by a Vickers indenter with a load 1.0 N (PMT-3M, Russia).

3. Results

X-ray patterns and phase identification are shown in Fig. 1. It is revealed that the matrix of the alloys were composed of austenite γ-phase (face centered cubic lattice, FCC)), which corresponds to the formation of invar alloy. Small peaks corresponding to body centered cubic lattice (BCC, α-phase) crystal structure appeared in the diffraction patterns for samples with Ni contain (30 – 35wt. %), probably due to relatively large size of powders; it coincides with the literature data [6]. The peak intensity corresponding to the BCC phase was decreased with increase of Ni content (Table 1). An extrapolation of this dependence to the zero intensity allows to obtaine a limit of existence BCC-phase, equals 34.8 %.

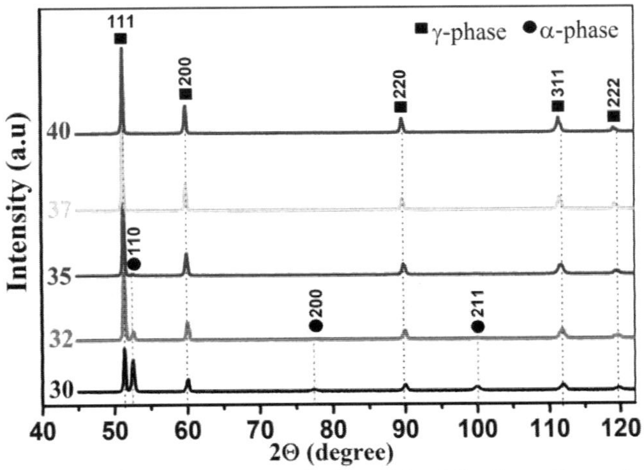

Figure 1. X-ray patterns and phase identification of samples

The structure of samples with 30 wt. % Ni (a) and 40 wt. % Ni (b) are shown in Fig. 2. As one can see clear austenitic microstructure with typical austenite twins was formed. The presence of α-phase is detected by light microscopy after etched of samples (Fig. 2, a – black area). The average grain size of the austenite decreases from 26.5±12.3 μm to 18.3±9.1 μm with increasing of Ni content from

30 wt. % to 40 wt. %. The porosity of samples increases from 13 ± 1.4 % to 18.4 ± 1.1 % with increasing Ni content.

Table 1. Phase composition of Fe-Ni compacts sintered at 1325 °C

The content in the mixture of Ni, wt. %	Relative content of phase, % (determined by summs of the integral line intensities)	
	FCC	BCC
30	63	37
32	90	10
35	98	2
37	100	0
40	100	0

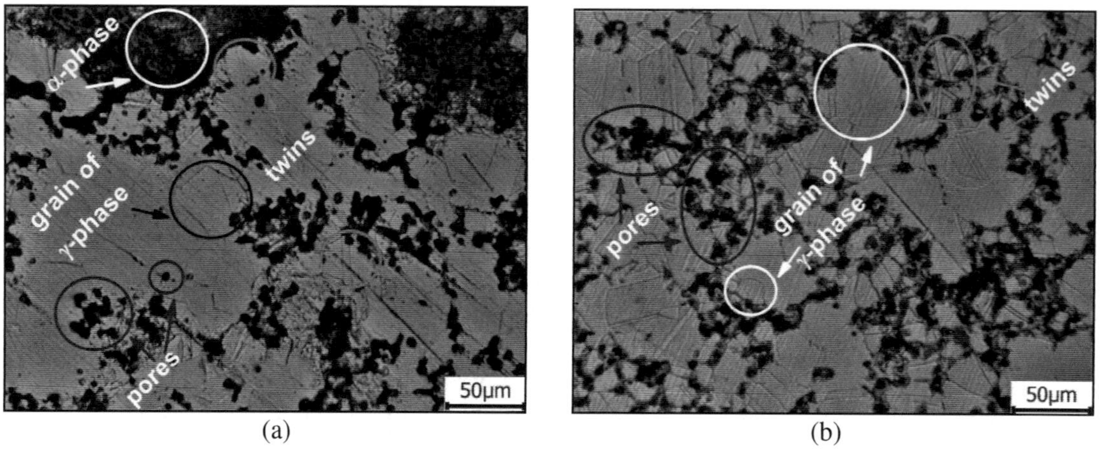

Figure 2. The structure of samples of Fe-30 wt. % Ni (a) and Fe-40 wt. % Ni (b)

The lattice parameters of γ-phase depending on the nickel content in sintered samples are shown in Fig. 3. One can see that the lattice parameters increase up to 37 wt. % of Ni and almost does not change for more content of Ni. It is in a good agreement with literature data [13].

The CDD of γ-phase is 50±5 nm and does not significantly change with increasing nickel content. The microdistortion of the crystal lattice of γ- phase changes from $1.9\pm0.15\cdot10^{-3}$ to $1.3\pm0.13\cdot10^{-3}$ with an increase of the nickel content, this may be associated with the disappearance of the α-phase in the alloy.

The dependence of microhardness (Hμ) of Invar alloys for different Ni content is shown in Fig. 4. It is shown, that H_μ of sintered samples changes non-monotonous for different Ni content, H_μ increases from 1.04±0.07 GPa to 0.78±0.05 GPa, after H_μ increases to 1.03±0.09 GPa. The first stage could be associated with α-phase; a second stage is connected with porosity and grain size of γ-phase. The H_μ value for Invar 36 [14] is 1.3 GPa and depends on a porosity and grain size of γ-phase for samples obtained by metal injection molding technology.

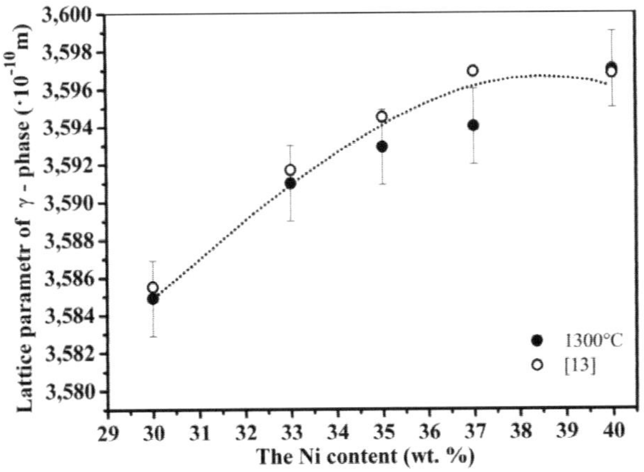

Figure 3. The lattice parameters of γ-phase depending on the nickel content in the initial mixture

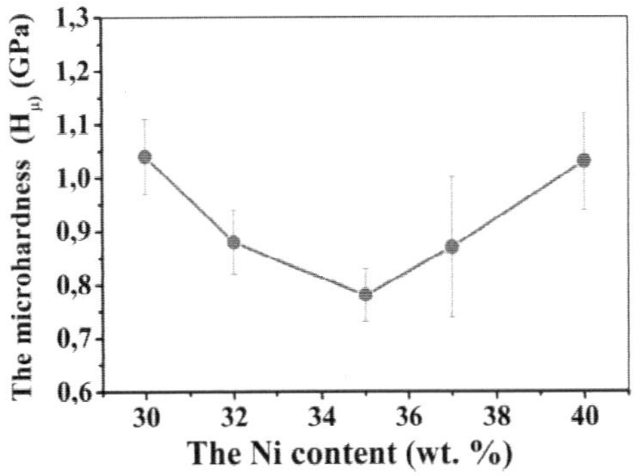

Figure 4. The dependence of microhardness (H_{μ}) of Invar alloys on Ni content

4. Conclusion

It has been shown that after sintering of Fe and Ni of powder mixtures were formed a face centered cubic structures with small amount of BCC-phase. Content of α-phase depends on Ni concentration and changes from 37 % to 2 %, and disappear at 34.8 wt. % of Ni. The average grain size of the γ-phase decreases with increase Ni content from 26.5±12.3 μm to 18.3±9.1 μm, and porosity of samples increases from 13±1.4 % to 18.4±1.1 %.The lattice parameter of the γ-phase depends on content of Ni and increases from $3.5849 \cdot 10^{-10}$ m to $3.597 \cdot 10^{-10}$ m which correlates well with literature data. It has been shown the microhardness of sintered samples changes non-monotonous with increase of Ni content, the maximum value is 1.04±0.07 GPa.

Acknowledgement
This research was funded by Russian Foundation for Basic Research and the government of the region of the Russian Federation, grant № 18-48-700039 and in the framework of the Program of increasing the competitiveness of TSU.

References

[1] Guillaume C E. 1897 *C R Acad Sci* **125**, 235–238.
[2] Maslyuk V A, Panasyuk O A, Vlasova O V. 2003 *Powder Metall. Met. Ceram.* **42** 536–539
[3] Lin C-T, Chiou B-S, Chi S. 2006 *IEEE J. Sel. Top. Quantum Electron.* **12(5)** 970–982.
[4] Cha B, Jang J M, Lee W et al. 2012 *J. Ceram. Process. Res.* **13**, 22–25.
[5] Duhamel C, Champion Y et al. 2005 *J. Alloys Compd.* **393**, 204–210.
[6] Hidalgo J, Jiménez-Morales A, Barriere T et al. 2014 *Powder Metall.* **2**, 127–136.
[7] Oglezneva S A, Saenkov K L, Grevnov L M. 2017 *Vestnik PNIPU* **3**, 34–48.
[8] Dovydenkov V A, Krys' M A, Fetisov G P. 2006 *Zagotovitel'nyye proizvodstva v mashinostroyenii.* **8**, 47–50.
[9] Luong D D, Shunmugasamy V C, Gupta N et al. 2015 *Mater. Des.* **66** 516–531
[10] Scherrer P. 1918 *Göttinger Nachrichten Gesellschaft.* **2** 98–101.
[11] Stokes A R, Wilson A J C. 1944 *Proceedings of the Physical Society.* 174–181.
[12] Pramanick A, Omar S, Nino J C et. al. 2009 *Appl. Crystallogr.* **42(3)** 490–495.
[13] Chamberod A, Laugier J, Penisson J M. 1979 *J. Magn. Magn. Mater* **10** 139–144.
[14] Cubberly W H, "Properties and selection – nonferrous alloys and pure metals", in ASM Metals

IOP Conf. Series: Materials Science and Engineering **696** (2019) 012006 doi:10.1088/1757-899X/696/1/012006

Recent advances in Bisphenol A UV/H_2O_2 degradation

E N Bocharnikova[1]*, O N Tchaikovskaya[1], V S Chaidonova[1], M. Gómez[2], M. Murcia[2], J. Gómez[2]

[1]Tomsk State University, 36, Lenin Ave., 634050, Tomsk, Russia;
[2]Universidad de Murcia, Campus de Espinardo, 30071, Murcia, Spain.

*E-mail: bocharnikova.2010@mail.ru

Abstract. Compounds destroying the endocrine system are a dangerous class of pollutants that are found in aquatic environments such as surface, groundwater and sewage and landfills. Bisphenol A (BPA) is a known endocrine disruptor that is acutely toxic to living organisms. BPA is still used in the manufacture of sunscreen lotions, nail polishes, body cleansers, shaving creams and lotions, in addition to its long-known other industrial applications. The paper presents the results of studies on the photolysis of BPA from water using ultraviolet radiation and an oxidizing agent.

1. Introduction

Microcontaminants more and more often fall into the focus of public discussion, especially such as endocrine drugs or other drugs that lead to various side effects for the elderly [1]. Pollution of surface and groundwater by objects of phenolic and other difficultly oxidized organic compounds occurs in the chemical, metallurgical, mining, pharmaceutical, pulp and paper, food industries and solid waste landfills.

Among synthetic xenoestrogens, bisphenol A (4,4'-(propane-2,2-diyl) diphenol) is one of the most common chemically synthesized compounds with annual production of over 3.8 million tons. BPA is widely used in the synthesis of polycarbonates, epoxies and thermal paper, in the manufacture of electronic equipment, paper or toys. BPA is also used in materials in contact with food. BPA is found in the wastewater of industries associated with the thermal processing of wood, shale, peat, lignite and coal, as well as due to their high foaming properties, they are widely used in flotation, etc. BPA disrupts the normal course of embryogenesis, leading to the appearance of various types of deformities in fish. As a result of the biochemical destruction of BPA in the water of reservoirs, all elements of the hydrochemical regime change.

One of the important tasks in modern society at the moment for the effective purification of liquid industrial effluents from phenolic compounds is the search for common patterns inherent in different schemes of transformation of substances, and the construction of models describing the effects of physico-chemical factors on the level of reduction of organic pollutants in the aquatic environment [2].

A modern and highly effective method for removing organic pollutants contained in industrial wastewater is the process based on the use of free radicals as oxidizing agents, called Advanced Oxidation Processes (AOP) [3], the advantage of which is the ease of controlling their speed in wide limits with a change in the intensity of the light flux, their rapid and complete suspension with the cessation of irradiation. It is known, •OH can be generated by irradiating water with UV light in the presence of oxidizing agents, such as hydrogen peroxide (H_2O_2). UV radiation is located at the high-energy end of the light spectrum with a wavelength less than that of visible light (400 nm), but more than that of x-rays (100 nm). UV radiation can destroy organic pollutants, including BPA, by direct and indirect photolysis.

Modern mercury-free excilamps have established themselves as efficient and environmentally friendly sources of quasi-monochromatic UV radiation for the decomposition of a number of organic pollutants (phenols, dyes, and herbicides) and are considered as an alternative to traditional mercury lamps [4].

The aim of this work is to study the effect of the irradiation wavelength on the efficiency of the phototransformation of BPA in water in the presence of H_2O_2 when irradiated with excilamps in a stationary KrCl and XeCl photoreactor.

2. Experimental part

In this work we used Bisphenol A (BPA), of chemical purity 95% (from ALDRICH) in water. Solutions of the investigated compound were prepared by dissolving a dry weighed quantity in the amount of 45.6 mg in 500 mL of water. To achieve complete solution of BPA to a concentration C =91.2 mg/L we used an ultrasound mixer at 40°C for 45 min. The structural formula of the investigated molecule is shown in Fig. 1.

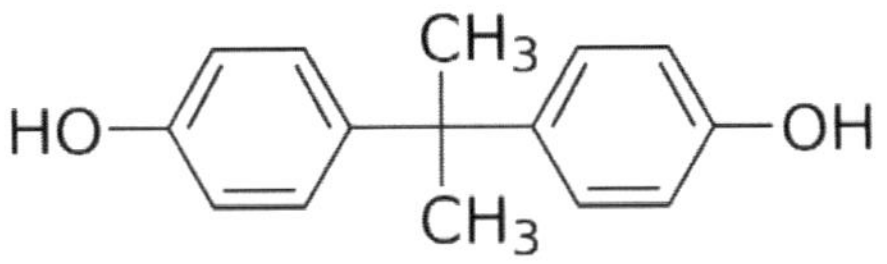

Figure 1. Structural formula of the investigated compound.

Spectral-luminescence characteristics of the solution of BPA were recorded before and after irradiation on a spectrophotometer "UV-Vis Spectrometry UNICAM" (Thermo Evolution 600, USA) and SOLAR CM2203 (Belarus). The prepared solutions of BPA with a concentration of 50 mg/L. were irradiated in glass beakers with a diameter of 4.6 cm by KrCl and XeBr excilamps at room temperature for 120 min with constant mixing with the help of a mechanical mixer. The distance from the excilamp to the irradiated solution was 3 cm (Fig. 2). The volume of irradiated solution was V = 50 mL. During the irradiation time the maximum energy absorbed by the investigated solution did not exceed 10 J/cm^3. The change in the volume and the temperature during irradiation was recorded. Depletion of the investigated compound was monitored by tracking variation of the optical density in the absorption spectra at the wavelength λ = 270 nm (the measurement error did not exceed 10%).

The conversion of BPA was calculated from values of the optical density at the maximum of the long-wavelength absorption band, according to the formula taken from [5-7]:

$$X = \frac{C_0 - C}{C_0} = 1 - \frac{C}{C_0}.$$

Designations:

V = volume of the solution during the irradiation (ml); V_0 = volume of the solution at the initial time of the irradiation (ml); C = concentration of BPA during the irradiation (mg/ml); C_0 = concentration of BPA at the initial time of the irradiation (mg/ml); t = exposure time (min); r = reaction rate (mg/ml min); λ_{irr} = maximum wavelength of excilamps (nm).

To prepare the systems with H_2O_2 we determined the concentration of the oxidizer according to the equation:

$$C_{14}H_{16}O_2 + X \times H_2O_2 = 14CO_2 + n \times H_2O.$$

In order to degrade 50 mg/L of BPA to CO_2 and water, it is necessary to have the maximum amount of H_2O_2 equal to 4.7 mg. The solutions were prepared from 34% of aqueous solution of H_2O_2. The conversion of BPA was calculated from values of the optical density at the maximum of the long-wavelength absorption band.

The photolysis model is based on following assumptions:

1. For photodegradation of BPA the oxidant H_2O_2 is required.
2. Photochemical reaction obeys the kinetics of a pseudo first order.

Material balance of BPA in the system of a stationary KrCl or XeBr photoreactor can be represented as follows:

$$Input = Output + Accumulation - Destruction.$$

The study of BPA photolysis was carried out simultaneously both in the absence and in the presence of hydrogen peroxide (Fig. 2). The reaction between •OH and many organic pollutants occur quickly; however, this reaction on its own account does not directly lead to the mineralization of these pollutants, but rather forms oxidation by-products, which can further react with •OH. Ideally, AOP systems are designed to completely mineralize organic pollutants to CO_2 and H_2O, but this needs to be optimized, as it can require a lot of energy and large doses of chemical oxidizing agents and, ultimately, can be expensive in some applications in purification systems.

We studied the effect of the irradiation wavelength on the effectiveness of the phototransformation of BPA in water in the presence of hydrogen peroxide. Irradiation of aqueous solutions was carried out at room temperature in a stationary photoreactor, the design of which is shown in Fig. 2. Constant mixing of the solution upon irradiation was included.

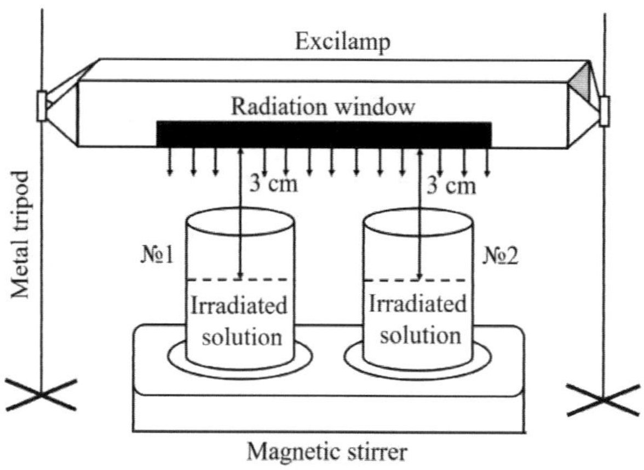

Figure 2. Scheme of a stationary photoreactor. The numbers indicate the solutions: without addition (№1) and in the presence of (№2) H_2O_2.

KrCl (with a radiation wavelength of 222 nm) and XeBr (with a radiation wavelength of 283 nm) were used as sources of UV irradiation. Static lamps had a cylindrical shape and were covered by a metal case with an output UV window of 75 cm^2. The distance from the source (excilamp) to the irradiated solution was 3 cm. The volume of the irradiated solution was 50 ml. The average intensity of radiation supplied to the solution was 17.12 and 2.47 mW/cm^2 for XeBr and KrCl excimer lamps, respectively. The control exposure time was: 0, 1, 2, 5, 10, 20, 30, 40, 60 and 120 minutes. The ratio of H_2O_2: BPA in the aqueous solution was as follows: 0:1, 1:1, 2:1, 3:1, 4:1. During the irradiation, samples were taken and absorption spectra were recorded for rapid visual observation of phototransformation and fixation of the concentration of total phenol in solution.

Analysis of the decrease in BPA in solution was also carried out by fluorescence. To construct the conversion curve, a decrease in the concentration of BPA in aqueous solutions was performed using HPLC.

3. Results and discussion

Taking into account that phenol compounds of various structures are expected to form during photodegradation of BPA, we analyzed the total phenol by electron spectroscopy. For this, the wavelength λ_{Dmax} = 740 nm was chosen, at which we measured the absorbance of the BPA solutions after irradiation. The data are shown in tables 1 and 2. Spectral interpretations allow a comparative assessment of the resulting photoproducts in irradiated solutions. A high content of common phenolic components was found. After irradiation, the optical absorption density increased in this region to a maximum value of 0.72 at a ratio of H_2O_2:BPA in an aqueous solution of 3:1 after 120 minutes of irradiation by XeBr excilamp. This indicates the active transformation of BPA and the formation of stable phenol-based photoproducts. The minimum absorbance value of 0.027 is achieved when the oxidizer to BPA ratio is 5:1 after 120 minutes of irradiation by KrCl excilamp. This suggests that with this ratio not only the destruction of BPA, but also of photo products occurs.

To construct the conversion curve, the decrease in the concentration of BPA in aqueous solutions was taken from the data using HPLC. An analysis of the rate constant of the decrease in BPA in water under the action of KrCl and XeBr radiation by excilamps indicates that, under the action of KrCl excilamp radiation, the efficiency of degradation of BPA without additives in water is higher than after exposure to XeBr excilamp [8]. In this case, the decrease in BPA in water after irradiation was only 60% after 120 min exposure to an excilamp (Fig. 3). The HPLC data of the final KrCl excilamp irradiated for 120 min in an aqueous solution showed that the solution also contains phototransformation products in addition to BPA.

Table 1. Absorption optical density of irradiated BPA solutions at a wavelength of 740 nm at various ratios of H_2O_2:BPA in an aqueous solution after irradiation of a KrCl excilamp using electron spectroscopy data.

t,	Mass ratio H_2O_2:BPA						
min	0:1	1:1	2:1	3:1	4:1	5:1	6:1
0	0.444	0.447	0.474	0.474	0.457	0.515	0.521
1	0.393	0.347	0.289	0.413	0.456	0.399	0.397
2	0.400	0.364	0.352	0.543	0.394	0.398	0.392
3	0.407	0.441	0.563	0.514	0.397	0.415	0.404
5	0.481	0.457	0.477	0.544	0.499	0.586	0.554
10	0.473	0.487	0.517	0.611	0.569	0.586	0.615
20	0.414	0.508	0.517	0.504	0.566	0.508	0.519
30	0.452	0.489	0.496	0.562	0.482	0.518	0.497
40	0.389	0.480	0.489	0.523	0.466	0.487	0.448
60	0.353	0.448	0.317	0.410	0.133	0.084	0.083
90	0.328	0.432	0.278	0.135	0.099	0.036	0.017
120	0.336	0.410	0.127	0.341	0.094	0.027	0.053

Table 2. Absorption optical density of irradiated BPA solutions at a wavelength of 740 nm at various ratios of H_2O_2:BPA in an aqueous solution after XeBr irradiation with an excilamp according to electron spectroscopy.

t,	Mass ratio H_2O_2:BPA						
min	0:1	1:1	2:1	3:1	4:1	5:1	6:1
0	0.337	0.458	0.488	0.497	0.459	0.474	0.479
1	0.381	0.480	0.487	0.416	0.363	0.623	0.370
2	0.415	0.475	0.523	0.512	0.455	0.423	0.392
3	0.413	0.456	0.457	0.588	0.353	0.555	0.371
5	0.368	0.352	0.544	0.446	0.469	0.394	0.394
10	0.415	0.460	0.460	0.498	0.473	0.538	0.523
20	0.377	0.454	0.489	0.574	0.502	0.559	0.555
30	0.381	0.474	0.499	0.568	0.527	0.599	0.572
40	0.369	0.464	0.537	0.622	0.555	0.692	0.605
60	0.341	0.465	0.499	0.634	0.581	0.541	0.551
90	0.384	0.421	0.477	0.662	0.532	0.422	0.468
120	0.363	0.418	0.510	0.720	0.496	0.395	0.383

The results of the photolysis of BPA in water with the addition of hydrogen peroxide are shown in Figure 3 When using the KrCl/H_2O_2 combination, the photolysis efficiency also increases in comparison with the XeBr/H_2O_2 combination.

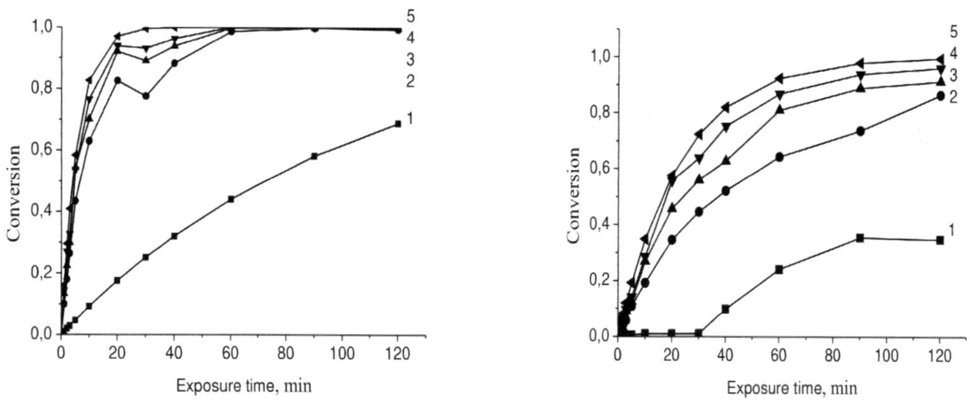

a *b*

Figure 3. The conversion of BPA in an aqueous solution in the presence of H_2O_2 at the ratio H_2O_2:BPA: 1 – 0:1; 2 – 1:1; 3 – 2:1; 4 – 3:1; 5 – 4:1. The irradiation was under the action of KrCl (a) and XeBr (b) excilamps.

The obtained curves of the dependence of the decay rate on time under the action of XeBr and KrCl excilamp radiation are shown in Figure 4.

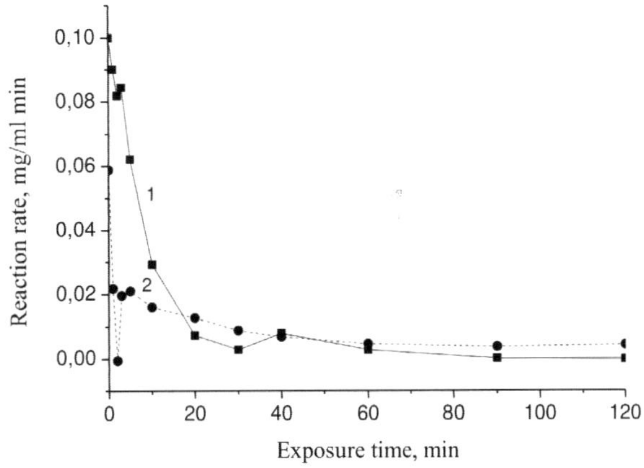

Figure 4. Decay rate of BPA in the water in relation to H_2O_2:BPA – 1:1 after the irradiation by XeBr excilamp (1) and KrCl excilamp (2).

An analysis of the nature of the obtained curves (Fig. 4) shows that the maximum rate when H2O2 is added to BPA solutions in water is observed upon exposure to KrCl radiation for a ratio of H2O2: BPA - 4:1 after 20 min of irradiation (Table 3). When exposed to XeBr radiation from the excilamp, the decrease in BPA at the same ratio with the oxidizing agent was achieved only after 120 min of irradiation (Table 4). Thus, the presence of H2O2 in the system contributes to an increase in the phototransformation efficiency of BPA under the influence of 222 nm irradiation.

The obtained HPLC data indicate that the irradiation wavelength changes the decay rate and the kinetics of the decomposition reaction of BPA photoproducts in the presence of an oxidizing agent.

Table 3. The conversion and concentration (C) of BPA in water after irradiation of KrCl excilamp according to HPLC.

t, min	H_2O_2:BPA ratio in aqueous solution									
	0:1		1:1		2:1		3:1		4:1	
	C, mg/l	Conversion	C, mg/l	Conversion	C, mg/l	Conversion	C, mg/l	Conversion	C, mg/l	Conversion
0	53.7	-0.039	52.4	-0.001	51.5	0.004	50.8	-0.011	49.4	0.021
1	52.4	-0.014	47.2	0.099	44.6	0.138	43.5	0.135	43.3	0.142
2	52	-0.006	43	0.179	39.9	0.228	37.1	0.262	34.9	0.309
3	49.4	0.045	38.6	0.263	36	0.303	34.2	0.319	24.3	0.520
5	50.9	0.015	29.7	0.433	23.5	0.545	23.6	0.531	22.4	0.556
10	46.7	0.097	19.5	0.629	15.4	0.702	11.9	0.763	10.3	0.795
20	44.3	0.144	9.2	0.825	4.1	0.920	3.1	0.938	2	0.960
30	35.5	0.314	11.8	0.775	5.7	0.890	3.6	0.929	0	1.000
40	34.1	0.341	6.2	0.882	3.2	0.938	1.9	0.962	0.7	0.986
60	30.7	0.406	0.8	0.985	0.1	0.997	0	1.000	0	1.000
90	27.3	0.472	0.2	0.996	0.1	0.998	0	1.000	0	1.000
120	10.4	0.799	0.4	0.992	0.2	0.997	0.1	0.998	0	1.000

Table 4. The conversion and concentration (C) of BPA in water after irradiation with XeBr excilamp according to HPLC.

t, min	H_2O_2:BPA ratio in aqueous solution									
	0:1		1:1		2:1		3:1		4:1	
	C, mg/l	Conversion	C, mg/l	Conversion	C, mg/l	Conversion	C, mg/l	Conversion	C, mg/l	Conversion
0	53.4	0.000	52.4	0.000	51.7	0.000	50.3	0.000	50.5	0.000
1	52.3	0.021	49.3	0.059	50.6	0.021	49.4	0.017	49.6	0.018
2	52.4	0.020	50.1	0.044	50.3	0.026	47.1	0.064	46.4	0.081
3	53.3	0.001	49.4	0.058	46.9	0.093	45.4	0.096	44.1	0.127
5	54.3	-0.017	46.7	0.108	45.2	0.126	43.1	0.142	40.7	0.194
10	50.6	0.052	42.3	0.192	37.8	0.270	35.9	0.287	32.2	0.363
20	49.8	0.067	34.3	0.346	28.1	0.457	22.3	0.557	21.1	0.582
30	49.4	0.074	29	0.446	22.7	0.560	18.1	0.639	13.7	0.728
40	47.1	0.118	25.1	0.522	19.2	0.629	12.4	0.753	8.8	0.825
60	40.3	0.245	18.7	0.644	9.8	0.810	6.6	0.868	5.8	0.886
90	38.4	0.281	13.8	0.737	5.8	0.889	3.1	0.938	1.9	0.962
120	37.7	0.294	7.1	0.864	4.5	0.912	2	0.960	0.7	0.987

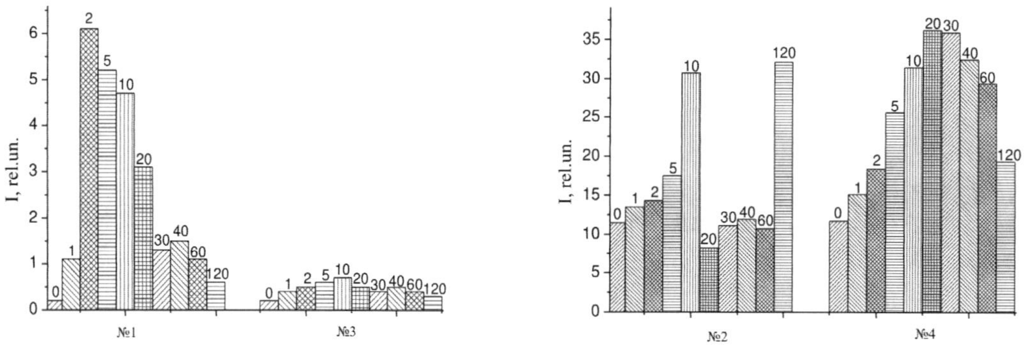

Figure 5. Diagrams of the formation of photoproducts in a photoreactor under the influence of excilamp radiation (№ 1 – KrCl, № 2 – KrCl/H_2O_2, № 3 – XeBr, № 4 – XeBr/H_2O_2) according to fluorescence spectroscopy. The numbers in the figure indicate the exposure time in the photoreactor.

The UV radiation source of the KrCl excilamp with an emission wavelength of 222 nm proved to be more efficient for direct photolysis of BPA compared to the XeBr excilamp with an emission wavelength of 283 nm. However, in all cases, BPA in water is more efficiently degraded with the addition of oxidation compared to direct UV treatment. Full mineralization was achieved only in the case of a combination of UV/H2O2 when exposed to KrCl radiation excilamps. However, according to fluorescence spectroscopy, it was found (Table 5) that even after 120 minutes, fluorescent photo products remain in the BPA solution. Moreover, the diagram clearly shows (Fig. 5) that during the phototransformation of BPA under the influence of KrCl radiation, the photo-product accumulates for 1 to 10 minutes, then the product breaks up and the fluorescence of photo-product 2 increases by 120 minutes. When exposed to XeBr radiation, the excilamp increases the fluorescence intensity of the photoproduct is 1 to 120 min of irradiation, and then the photoproduct is transformed.

Table 5. The fluorescence intensities of the resulting photoproducts in the maximum bands.

№	Excilamp	The maximum fluorescence band of the photoproduct, nm	Exposure time, min									
			0	1	2	5	10	20	30	40	60	120
1	KrCl	λ=410-420	0.2	1.1	6.1	5.2	4.7	3.1	1.3	1.5	1.1	0.6
2	KrCl/H_2O_2	λ=430-450	11.5	13.5	14.3	17.5	30.7	8.2	11.1	11.9	10.7	32.1
3	XeBr	λ=410-420	0.2	0.4	0.5	0.6	0.7	0.5	0.4	0.5	0.4	0.3
4	XeBr/H_2O_2	λ=430-450	11.7	15.1	18.4	25.6	31.4	36.2	35.9	32.4	29.4	19.3

4. Conclusion

Prior to our studies, an assessment of human exposure to BPA was established. It has been proven that BPA can affect animals and humans when interacting with estrogens, endogenies and receptors, disrupting the function of not only the endocrine system, but also the immune or nervous system. The results of our studies showed that photodegradation of BPA in water in the presence of an oxidizing agent can be successfully used to remove the contaminant before entering into natural waters in order to reduce the risk to living organisms. We will have to further study the toxicity of BPA photolysis photoproducts and revise the kinetic model of contaminant degradation in the presence of an oxidizing agent. Perhaps the decay of the BPA will be subject to a decay curve of pseudo-second order.

Acknowledgments

The research was supported by Russian Ministry of Science and Education at The Tomsk State University competitiveness improvement programme (project number 8.1.20.2018).

References

[1] Daverio E, Ghiani M, Bernasconi C 2013 *Ecotoxicol Environ Saf.* Vol **91(13)** Institute for Environment and sustainability, Joint Research Center, European Commission p 1-9.

[2] Karmazinov F V, Kostyuchenko S V, Kudryavtsev N N, Khramenkov S V 2012 *Publishing House Intellect* Dolgoprudny p 392.

[3] Sokolov E M, Sheinkman L E, Dergunov D V 2013 *Bulletin of the Southern Scientific Center of the Russian Academy of Sciences* Vol **9(2)** Tula Russia p 23-31.

[4] Parrott J L, Blunt B R. 2005 *Environmental Toxicology* Vol **20** Burlington Ontario Canada p 131-141.

[5] Tchaykovskaya O N, Bryantseva N G, Bocharnikova E N, Charushin V N 2017 *Proc. SPIE 10614* doi:10.1117/12.2303512.

[6] Tchaikovskaya O, Bryantseva N, Krayukhina V 2017 *Proc. SPIE 10614* doi:10.1117/12.2302665.

[7] Gómez M, Murcia M D, Gomez E, Ortega S, Sanchez A, Tchaikovskaya O N, Bryantseva N G 2016 *Journal of Environmental Management* Vol **183(3)** Murcia Spain p 818-825.

[8] Gómez M, Murcia M D, Gomez E, Tchaikovskaya O N, Bocharnikova E N, Krayukhina V S 2019 *ANQUE-ICCE-CICIQ* Santander Spain.

IOP Conf. Series: Materials Science and Engineering **696** (2019) 012007 doi:10.1088/1757-899X/696/1/012007

Comparative assessment of characteristics of industrial and laboratory highly-efficient samples of adsorbents-desiccants

A V Livanova, M P Sandu, Zh B Budaev, E P Meshcheryakov

Tomsk State University, Russia, Tomsk, Lenin ave., 36, 634050

E-mail: truelivanova@mail.ru

Abstract. The research results of the physicochemical characteristics, adsorption kinetics and the value of adsorptive capacity, with reference to water vapors, of industrial adsorbents-desiccants Alumac 2-5H, Alusorb 675 and experimental samples based on aluminum oxide, obtained by centrifugal thermal activation of hydrargillite with its subsequent hydration in mild conditions, are presented. It is demonstrated that modification by sodium cations during synthesis at the hydration stage or impregnation with hygroscopic salt ($CaCl_2$) of the granules of synthesized aluminum oxide adsorbent allows obtaining desiccants with a higher value of dynamic capacity with respect to water vapors as compared to industrial adsorbents. The most balanced set of characteristics belongs to the aluminum oxide sample containing 0.6 % mass. of sodium, for which static capacity by water is higher than 20 g/100 g; the dynamic capacity value is at a level of 6.9 g/100 cm^3, and the mechanical strength is 8.4 MPa.

1. Introduction

Aluminum oxide possesses unique properties; therefore, it is applied in different areas of production, in particular, as an adsorbent of water vapors from moisture-containing gases [1]. Water resistance is an important positive property of aluminum oxide and it often determines the selection of aluminum oxide as an adsorbent for drying and treatment of media containing condensed moisture. The use of a more effective desiccant (Al_2O_3) as a protective layer will allow for not only the protection of a bottom layer from condensed moisture but also an enhancement of the effectiveness and operational life of the adsorber [2]. The peculiarity of the developed adsorbents-desiccants based on aluminum oxide is their different phase and chemical compositions [3,4]. This work presents the research of the physicochemical and adsorption properties of two laboratory samples of adsorbents based on aluminum oxide, one of which is modified by sodium cations [3], and the other one is covered with hygroscopic salt – $CaCl_2$ [4] as compared with the samples of industrial aluminum oxide adsorbents in the comparable conditions.

2. Materials and methods of research

Laboratory samples of aluminum oxide desiccants based on bayerite and pseudoboehmite, synthesized by the technology of centrifugal and thermal activation of hydrargillite (CTA HG) with its subsequent hydration in mild conditions were used in the work [5,6]. Modification by sodium cations (sample OA-Na) was implemented at the stage of hydration of products of thermal activation of hydrargillite during synthesis of the bayerite-containing adsorbent. The pseudoboehmite-based (OA-$CaCl_2$) adsorbent sample impregnated with hygroscopic salt $CaCl_2$ was obtained by impregnating the matrix from the excess solution (equilibrium deposition filtration) [7]. A γ-Al_2O_3 sample in the form of granules of cylinder shape (5.0÷6.0 mm in length, 3.5÷3.6 mm in diameter) was used as a matrix (OA). The laboratory samples were compared with the industrial samples based on activated alumina of Alumac 2-5H (AXENS Company, France) and Alusorb 675 (Salavat Catalyst Plant, Russia) grades, which were used for gas dewatering and cleaning from foreign and chemical impurities.

The mass fraction of Ca or Na in synthesized sorbents was determined by atomic emission spectroscopy of microwave plasma using the atomic emission spectrometer of microwave plasma "Agilent 4100" (Agilent -Technologies, USA) with preliminary "disclosure" of acids in the mixture in the system of microwave sample preparation "Speedwave four SW-4" (Berghof, USA). The content of the $CaCl_2$ salt in the adsorbent was calculated proceeding from the metal content. The textural characteristics of the adsorbents were determined by isotherms of nitrogen adsorption-desorption at 77 K using the 3Flex sorptometer (Micromeritics, USA). The mechanical strength of granules was assessed on the device "IPG -100" (Federal State Unitary Enterprise "UNIHIM with OZ", Russia). The measurements were taken for 30 granules and the average value was found. The dynamic adsorption capacity (DC) of adsorbents with reference to water vapors was determined by means of measuring the amount of water vapors adsorbed from their mixture with the air, passing through the layer of adsorbent granules in the adsorber, at the moment of reaching the dew point temperature of the gas flow on exit from the adsorber (minus 40.0 °C). The resulting value was related to 100.0 cm^3 of the test sample. The layer height of the test sample (with the granule sizes of 2.0-4.0 mm in diameter and 2.0-4.0 mm in length) was 28.0-29.0 cm. The static adsorption capacity of the adsorbents with respect to water vapors was measured using the gravimetric method in the exsiccator. The static adsorption capacity (SC) of the samples was determined by the mass fraction of water absorbed by a mass unit of the adsorbent by the time of reaching adsorption equilibrium at a temperature of 25.0 °C in the atmosphere with relative humidity of 60.0 %. Experiments on the study of adsorption kinetics of water vapors on the modified samples were conducted using the adsorption plant and the quartz balance at a temperature of 25.0 °C and relative humidity of 100 % according to the procedure, described in work [8].

3. Results

According to the results of X-ray diffraction analysis, shown in Figure 1, the major phase of the samples under study is γ-Al_2O_3. The boehmite phase is present in the X-ray patterns of the industrial samples Alumac 2-5 H and Alusorb 675. For the OA-$CaCl_2$ sample, the calcium chloride phase is not identified probably because of its X-ray amorphousness. The assumed interval of peaks of the calcium chloride phase is 15-40 2θ.

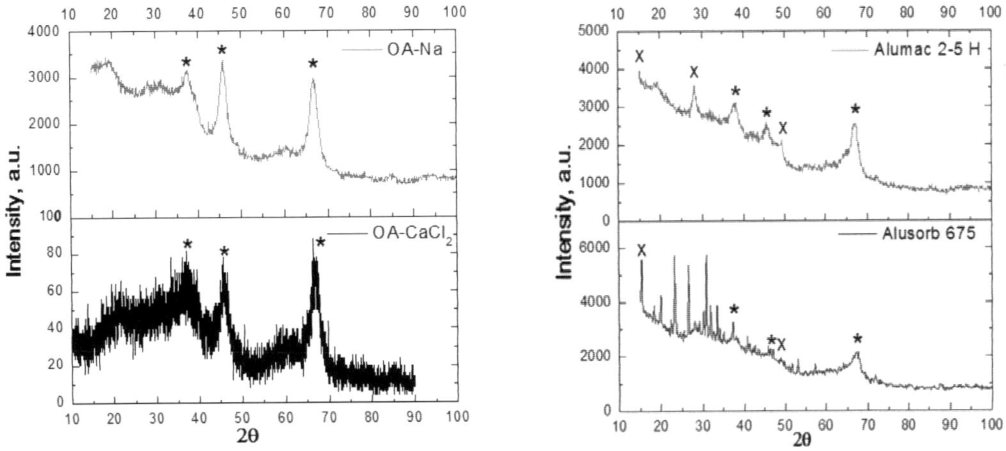

Figure 1. XRD-analysis of aluminium oxide materials (* - γ-Al_2O_3, x - boehmite).

As a result of studying the adsorption process, it was established that isotherms of nitrogen adsorption at relative pressures of 0.05-0.2 on all the samples were described by the BET equation. These isotherms belonged to isotherms of the IV type according to the IUPAC classification with an initially sudden rise at a low relative pressure, which was conditioned by the presence of micropores and a gradual increase in the adsorption value with a pressure increase. In the range of relative pressures over 40 %, adsorption-desorption isotherms exhibited pronounced hysteresis. This type of the isotherm pointed to the presence of mesopores on the surface of the samples under study, as well as the presence of reversible capillary condensation in adsorbent mesopores. The isotherms of low-temperature nitrogen sorption of industrial and modified laboratory samples of adsorbents are presented in Figure 2. It is established that the larger adsorption capacity by nitrogen belongs to the industrial samples of aluminum oxide adsorbents.

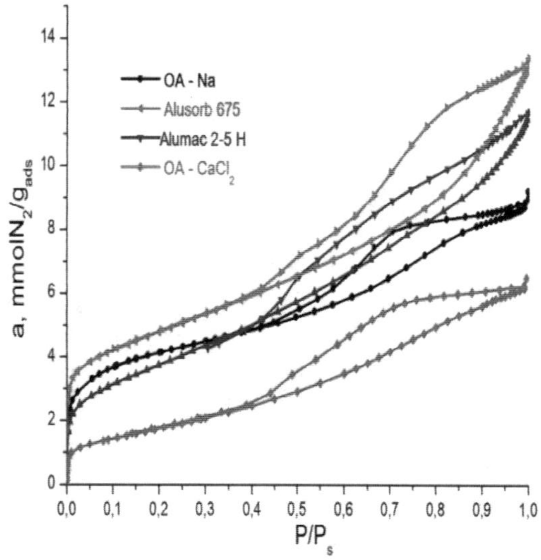

Figure 2. Isotherms of nitrogen adsorption-desorption on the samples.

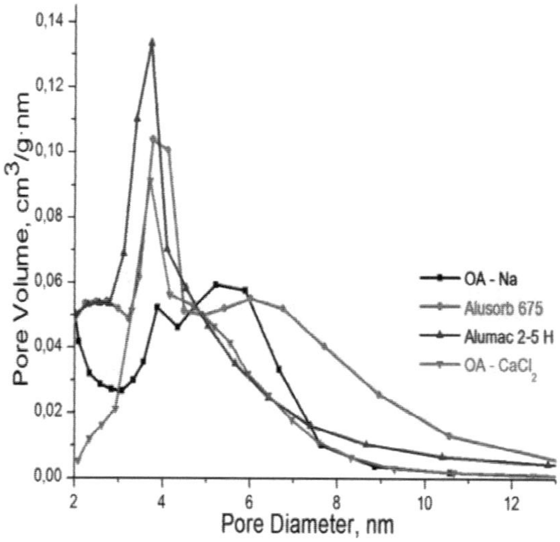

Figure 3. Pore size distribution of samples.

The composition, textural characteristics and values of the mechanical strength of the examined samples of the aluminum oxide adsorbents are given in Table 1.

Table 1. Textural characteristics, crushing strength in the aluminum oxide samples.

Sample	Composition of the sample	Shape and sizes of granule, mm	S_{BET} (m²/g)	V (cm³/g)	d (nm)	Micropore surface area (m²/g)	Limiting micropore volume (cm³/g)	Crushing strength (MPa)
Alumac 2-5H	aluminum oxide, boehmite	Sphere Ø = 2.0 ÷ 5.0	350	0.40	5.2	2	0.050	2.3
Alusorb 675	Al_2O_3, boehmite	Sphere Ø = 4.0 ÷ 4.3	373	0.45	4.9	77	0.037	14.5
OA	Al_2O_3	Cylinder 3.6x5.2	290	0.36	4.8	226	0.104	7.1
OA-Na	Al_2O_3+0.6 mass. % Na	Cylinder 3.6x5.2	315	0.30	3.8	97	0.047	8.4
OA-$CaCl_2$	Al_2O_3+13.8 mass. % $CaCl_2$	Cylinder 3.6x5.2	144	0.21	5.9	119	0.055	4.9

The textural characteristics of the samples of the aluminum oxide were compered. It was established that the industrial samples of the adsorbents, initial sample and OA-Na sample have similar values of the specific surface, ranging from 290 to 380 m²/g. At the same time, the surface area, occupied by the micropores, and the micropore volume for the industrial adsorbents are lower. The value of the total volume and the average pore size for the industrial adsorbents is slightly higher than that of the initial sample of the aluminum oxide adsorbent and sample, modified by sodium cations. The introduction of hygroscopic salt into aluminum oxide makes major changes in the properties of hygroscopic salts and the initial adsorbent used as a matrix. This is manifested through the change in textural, adsorption and strength characteristics. When the salt content in aluminum oxide increases, the nitrogen adsorption capacity, volume of meso- and micropores, specific surface area decrease along with the increase in the diameter of mesopores. The pore-size distribution for all the examined samples of the desiccants is given in Figure 3. The samples under study are characterized by the presence in the structure of fine mesopores, being in the distribution interval of 3-8 nm. Samples OA-$CaCl_2$ and Alumac 2-5H have one maximum on the curve of pore-size distribution in a range of 3.5-4.3 nm. As regards Alusorb 675 and OA-Na samples, there is a bimodal distribution with the first maximum in the range of 3.5-4.3 nm and the second one in a range of 4.5-8.0 nm. The mechanical strength of the Alusorb 675 sample is the highest amongst all the analyzed adsorbents, and that of the foreign sample Alumac 2-5H is the lowest. The introduction of sodium cations into the adsorbent allowed increasing the mechanical strength of the OA-Na sample, and impregnation of the matrix with hygroscopic salt decreased the strength of the OA-$CaCl_2$ sample as compared to the initial sample.

The kinetic curves of adsorption of water vapors on the obtained samples of modified adsorbents are presented in Figure 4. The adsorption equilibrium for the OA-Na sample was reached in 90 minutes after the beginning of the experiment, and for the industrial samples – in 200-250 minutes. As for the sample, modified by calcium chloride, the water vapor adsorption capacity did not reach the equilibrium value after 5 hours since the beginning of the experiment.

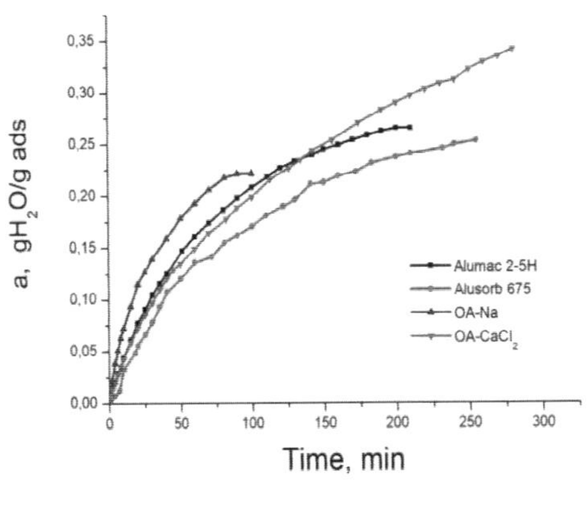

Figure 4. Kinetic curves of adsorption of water vapors on the samples for a fraction of 0.5–1.0 mm (conditions: carrier gas adsorption rate – 30 l/h).

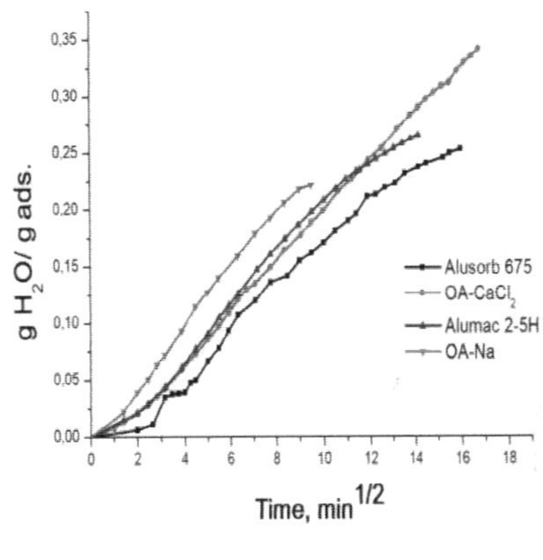

Figure 5. Approximation of experimental data by equation $a=A\,\tau^{1/2}$.

Approximation of the experimental data by different kinetic equations showed that the adsorption rate on all the samples was best described by the equation $a=A\,\tau^{1/2}$ (Figure 5), where a - the amount of water adsorbed at time τ; A - the rate constant, g/min$^{-1/2}$; τ - time, min. The values of rate constants A, determined in this equation for the samples under study, are given in Table 2. The equation of linear dependence of the adsorption value on $\sqrt{\tau}$ was earlier used specifically in the description of the kinetics of water vapor adsorption on aluminum oxide [9] and the composite material "calcium chloride/silica gel" [10].The equation was evidence of the fact that the limiting stage of the process is water diffusion in the adsorbent pores.

Table 2. Characteristics of the kinetic curves of water vapor adsorption on aluminum oxide samples and the values of static and dynamic capacity.

Sample	a_{max}, g_{water}/g_{ads}	$a=A*\tau^{1/2}$		SC (g/100g $_{ads}$) with 60% humidity	DC (g/100 cm^3)
		A (g /min$^{-1/2}$)	R[b]		
Alumac 2-5H	0.27	0.0208	0.99	21.1	4.4
Alusorb 675	0.25	0.018	0.98	21.0	5.0
OA-Na	0.22	0.025	0.99	20.3	6.9
OA-$CaCl_2$	0.34[a]	0.022	1.00	23.2	5.8
OA	0.25	0.020	0.99	23.4	4.6

[a] value of the adsorption capacity in 280 minutes after the beginning of the experiment;
[b] R – linear correlation coefficient.

The values of the ultimate adsorption capacity by water, determined in the kinetic experiments, decrease amongst Alumac 2-5H, Alusorb 675, OA-Na. The values of static capacity change symbatically for these samples as well. Dynamic capacity in passing from Alumac 2-5H to OA-Na increases. As to the sample impregnated with calcium chloride, in a time of conducting the experiment (280 minutes), the value of the ultimate adsorption capacity was not obtained (Figure 4). A large value of static capacity (Table 2) and higher as compared to the industrial adsorbents dynamic capacity turned out to be typical of this sample. However, the largest dynamic capacity and adsorption rate among all the studied samples belong to the laboratory sample modified by sodium cations (OA-Na).

4. Conclusions

The introduction of modifying additives into aluminum oxide changes significantly the adsorbent properties. This is manifested in the change of textural, adsorption and strength characteristics. When aluminum oxide is impregnated with hygroscopic salt, the value of the specific surface decreases, the volume of meso- and micropores reduces. In this case the primary contribution into the adsorption process is made by the absorption of water vapors by hygroscopic salt [4]. As a result of the alkaline modification, the specific surface increases, but the average diameter and pore volume decrease. The most probable reasons for increasing the sorption indicators of this adsorbent, as compared to the initial sample (OA), can be a higher specific surface (Table 2), as well as formation of optimal acid-base properties of the surface, determined by the chemical modification. The kinetic curves of water vapors adsorption for all the adsorbents are best described by the equation applied for the processes the rate of which is limited by diffusion in the adsorbents pores. Adsorption rate constant A for the modified samples is higher than that of the initial sample and samples of industrial adsorbents.

All the examined samples of the adsorbents are characterized by the presence in the structure of small mesopores, pertaining to a narrow distribution range of 3.5-8.0 nm. The value of the specific surface for the industrial samples of adsorbents is slightly larger than that for the laboratory samples; however, the surface area, occupied by micropores, is less. In the micropores (with a diameter of less than 2 nm), there is the volume filling of the adsorption space. On the one hand, mesopores transport water vapors to micropores; on the other hand, mesopores adsorb water by means of the capillary condensation. When dewatering different gases in industry, dynamic sorption is most often applied. Therefore, the value of dynamic capacity of the desiccant, depending on the value of its specific surface, the volume of capacity and transport pores and their ratios, is of paramount importance when selecting an adsorbent. The conducted research demonstrates that the obtained experimental samples surpass the samples of the industrial adsorbents by this parameter. Besides, they have quite high static capacity and mechanical strength. The aluminum oxide sample, modified by sodium cations, turns out to be the most promising for using as a desiccant throughout the entire set of characteristics. For this sample, static capacity by water is higher than 20g/100g, the value of dynamic capacity is at a level of 6.9 g/100 cm^3, and the mechanical strength is 8.4 MPa.

Acknowledgments
The study was financially supported by project RFMEFI57517X0139 of the Ministry of science and higher education of the Russian Federation.

References

[1] Ivanova A 2009 *Promyshlennyi kataliz v lekciyah* **8** 7–9
[2] Zotov R, Meshcheryakov E, Livanova A, Minakova T, Magaev O, Isupova L and Kurzina I. 2018 *Materials* **11** 132 DOI: 10.3390/ma11010132
[3] Livanova A V, Meshcheryakov E P and Kurzina I A 2019 Synthesis of aluminum-oxide adsorbents using alkalis at the hydration stage and study of their physicochemical and adsorptive properties *Materials conference of PFSD-2019* 126-129
[4] Budaev Zh B, Livanova A V, Magaev O V 2019 Influence of the composition of the adsorbent "hygroscopic salt/aluminum oxide" on its physicochemical properties *Materials conference of PFSD-2019* 42-45
[5] Danilevich V V, Isupova L A; Paukshtis E A and Ushakov V A 2014 *Kinet. Catal.* **55** 372–79
[6] Danilevich V V, Isupova L A, Kagyrmanova A P, Kharina I V, Zyuzin D A, Noskov A S 2012 *Kinetics and Catalysis* **53** 673-80
[7] Bourikas K, Kordulis C and Lycourghiotis A 2006 *Catal. Rev.* **48** 363-444
[8] Reshetnikov S I, Livanova A V, Meshcheryakov E P, Kurzina I A and Isupova L A 2017 *Russian journal of applied chemistry* **90** 1760-65
[9] Marcussen L 1974 *Chem. Eng. Sci.* **29** 206
[10] Ovoshchnikov D S, Glaznev I S and Aristov Yu I 2011 Kinetics and Catalysis 52 620-28

IOP Conf. Series: Materials Science and Engineering **696** (2019) 012008 doi:10.1088/1757-899X/696/1/012008

High-Temperature Laminated Ceramic Material Hfc-ZrB_2-SiC-ZrO_2 for the Development of the Arctic and Siberia

A G Burlachenko[2], Yu A Mirovoy[2], S P Buyakova[1,2], E S Dedova[2] and S N Kulkov[1,2]
[1] National Research Tomsk State University, Tomsk, 634050, Russia
[2] Institute of Strength Physics and Materials Science SB RAS, Tomsk, 634055, Russia

E-mail: burlachenkoag8@gmail.com

Abstract. Ultra-high-temperature ceramics based on zirconium diboride and silicon and hafnium carbides are presented as composites with ceramic matrix and oxidation-resistant surface. Relatively low density, high strength at extreme temperatures and operability at oxidative environment with temperature up to 2000°C make these materials useful for aerospace construction and power engineering. HfC-ZrB_2-SiC-ZrO_2 systems are presented, structural analysis and the results of temperature tests in high-enthalpy gas are provided. It was demonstrated that high resistance of HfC-ZrB_2-SiC-ZrO_2 system to high-temperature corrosion is achieved owing to formation of protective oxide layer with variable composition. A key area for the use of the material in the Arctic is the organization of heat-shielding structures in the oil and gas industry.

1. Introduction
One of the most promising aims of modern materials science is development of composite materials demonstrating resistance to high-temperature heat flows and having operation temperatures up to 2500°C [1–3]. Among the promising high-temperature composites, oxygen-free ceramics deserved significant attention. Their advantages are low density, outstanding operating temperature, chemical inertness, corrosion and crack resistance, high strength, etc. [2–4]. ZrB_2-based ceramics are the most interesting amongst high-melting materials. SiC-doped ZrB_2 ceramics demonstrate high thermal and chemical stability, in particular – oxidative resistance in extreme temperatures [1–6]. In last decades, a large amount of studies is focused at these high-melting compounds. However, composite functional-graded materials which possess various functional properties are of great interest. The present article describes the results of fabrication of heat-protective multilayered ceramic material with hierarchical structure organized by reduction of heat conductivity from top to bottom side. Such principal provides high heat insulation in extreme conditions. ZrB_2-SiC compounds were selected as a basic material for the composite. HfC coating was applied as a layer with the highest melting temperature (3890 °C). ZrO_2 was chosen for bottom layer as a material with abnormally low heat conductivity. Combination of properties of the mentioned compounds in multilayered material makes it promising in the sphere of aerospace construction.

2. Materials and methods
ZrO_2 (4 wt. % of Y_2O_3), ZrB_2 and SiC powders were used as primary components. Application of HfC coating was conducted using magnetron sputtering of Hf target in methane atmosphere.

Multilayered ceramic composites were obtained by hot-pressing at a temperature of 1800 °C and pressure of 35 MPa. Duration of isothermal treatment was 30 minutes. Phase composition and crystal structure parameters were studied using X-ray diffraction with CuK_α irradiation, 0.05° scanning step and 5 seconds exposure. Structure of the obtained materials was studied using Tescan VEGA 3 scanning electron microscope. Thermal treatment of HfC-ZrB_2-SiC-ZrO_2 composite materials for studies of oxidation and ablation resistance in high-enthalpy flow of dissociated air was conducted in Plasmotron EDG-200M apparatus.

Strength of the obtained materials was determined by diametral compression method using Instron test machine. Tensile strength under diametral compression (rupture) was calculated according to the following formula:

$$\sigma = \frac{0,636P}{d*h} \tag{1}$$

where: P – maximum load, d – sample diameter, h – thickness.

3. Results and discussion

Obtained HfC-ZrB_2-SiC-ZrO_2 multilayered composite is shown in Fig. 1. The layers are deposited on the principle of heat conductivity lowering from the top to bottom one with lowest gradient of thermal expansion coefficient. Top side of the sample is presented by ZrB_2-SiC layer with HfC coating, the bottom one is made of ZrO_2.

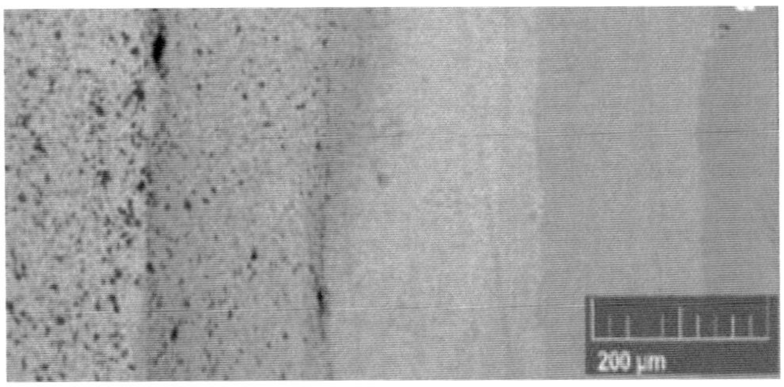

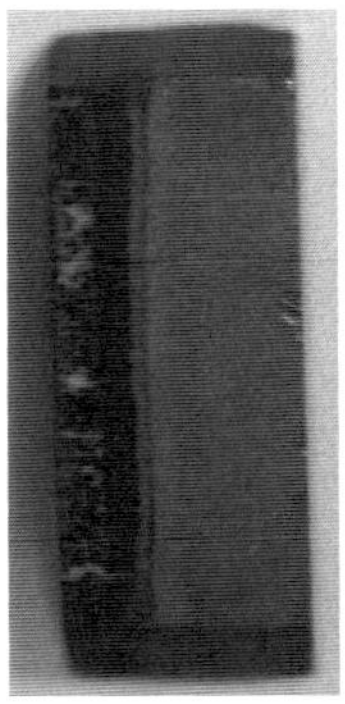

(a) (b)

Figure 1. Structure of the multilayered HfC-ZrB_2-SiC-ZrO_2 composite material (a); macro photo of the sample (b).

Table 1. Composition and temperature characteristics of material layers.

Layer Composition	α_L, $10^{-6}\times K^{-1}$	λ, W/m×K
Front layer HfC	6.1	17
ZrB_2-SiC	5.2	65
Rear layer, ZrO_2	10.4	2.3

SEM images of ZrB_2 – SiC layer with 20 vol.% of SiC coated with HfC and bottom ZrO_2 layer of HfC-ZrB_2-SiC-ZrO_2 system are shown in Fig. 2.

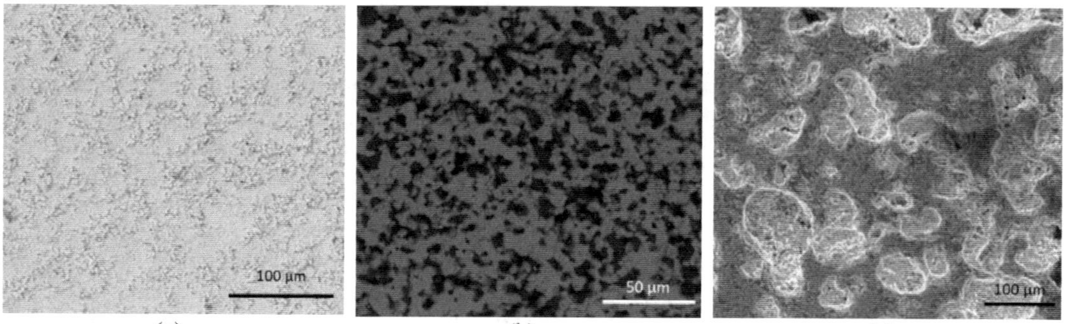

(a) (b) (c)

Figure 2. Images of the surface structure of the multilayer composite: a) Front layer HfC b) ZrB_2-SiC layer, c) rear layer.

Analysis of the structural state of the studied composites revealed that obtained HfC-ZrB_2-SiC-ZrO_2 materials possess homogenous distribution of SiC particles in ZrB_2-SiC layer. HfC coating on the ZrB_2-SiC layer was found uniform on all samples. XRD patterns of the top layer of HfC-ZrB_2-SiC-ZrO_2 heat-protective material with HfC coating are provided in Fig. 3.

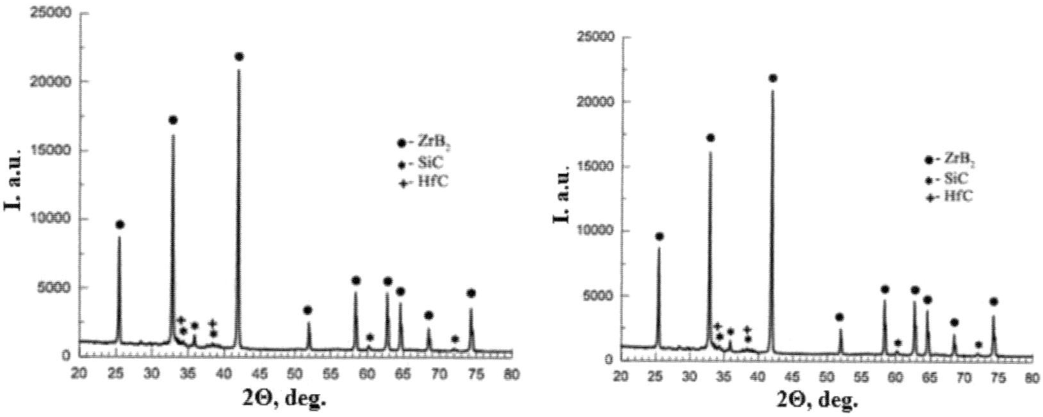

(a) (b)

Figure 3. XRD patterns of the HfC-coated samples of HfC-ZrB_2-SiC-ZrO_2 composite front layer with SiC concentration of 15 (a) and 20 (b) % v/v.

Phase analysis of the XRD patterns of all samples showed the presence of ZrB_2 reflexes. Moreover, low-intensity peaks corresponding to SiC were observed. HfC is presented by broad reflexes. That fact demonstrates nanostructural nature of the obtained coating.

Samples of HfC-ZrB_2-SiC-ZrO_2 material before and after 1000 seconds of high-enthalpy dissociated airflow exposure at a temperature of 2500°C are shown in Fig. 4. Formation of glasslike phase with ZrO_2 particles was mainly observed. This is due to thermal-oxidative reactions with oxygen from high-temperature plasma. Surface structure is characterized by heterogeneous distribution of oxidation products. Mass increase after high-temperature exposure was found at 0.3 g, crater depth was found at +0.13 mm.

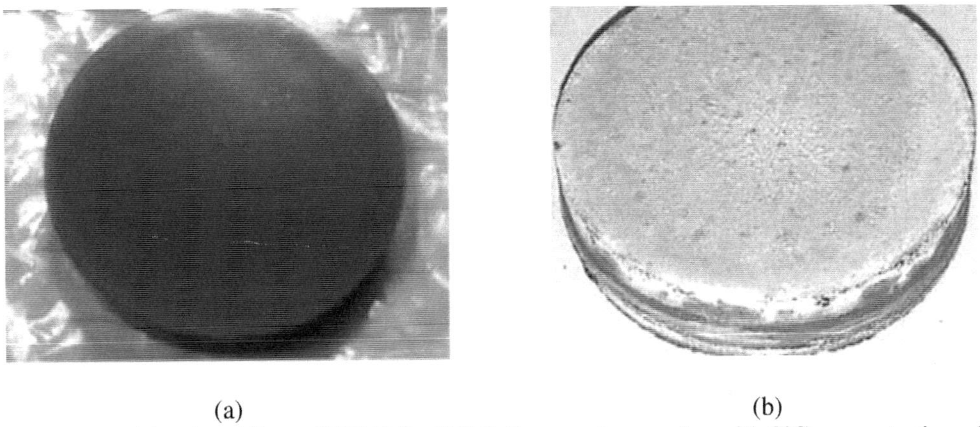

(a) (b)

Figure 4. Top side of multilayer HfC-ZrB_2-SiC-ZrO_2 ceramic samples with SiC concentration of 20 % v/v: a) before and b) after high-temperature testing.

Fig. 5 represents XRD patterns of the surface of multilayered heat-protective HfC-ZrB_2-SiC-ZrO_2 ceramics with heat-resistant HfC coating after high-temperature testing. Reflexes corresponding to ZrO_2 monocline modification were observed. HfO reflexes possess low intensity.

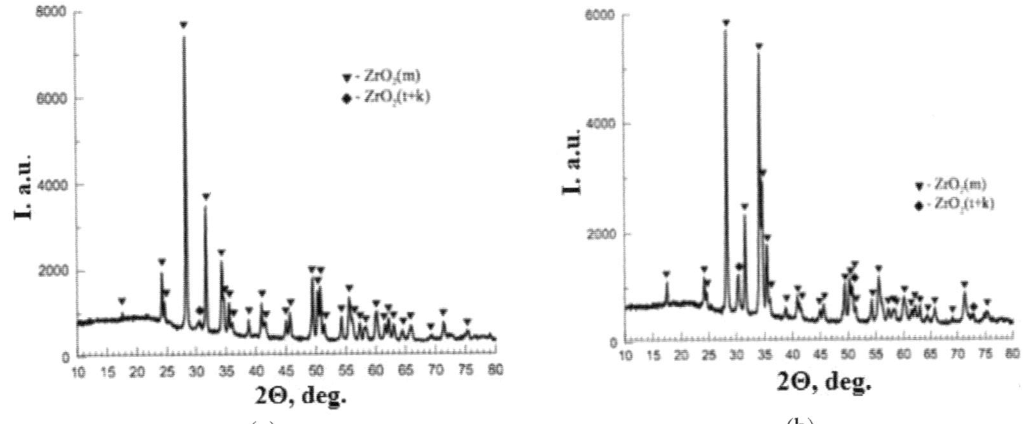

(a) (b)

Figure 5. XRD patterns of the samples after high-temperature testing with SiC concentration of 15 (a) and 20 (b) % v/v.

SEM images of the layers with SiC concentration of 15 and 20 % v/v after high-temperature testing are provided in Figure 6.

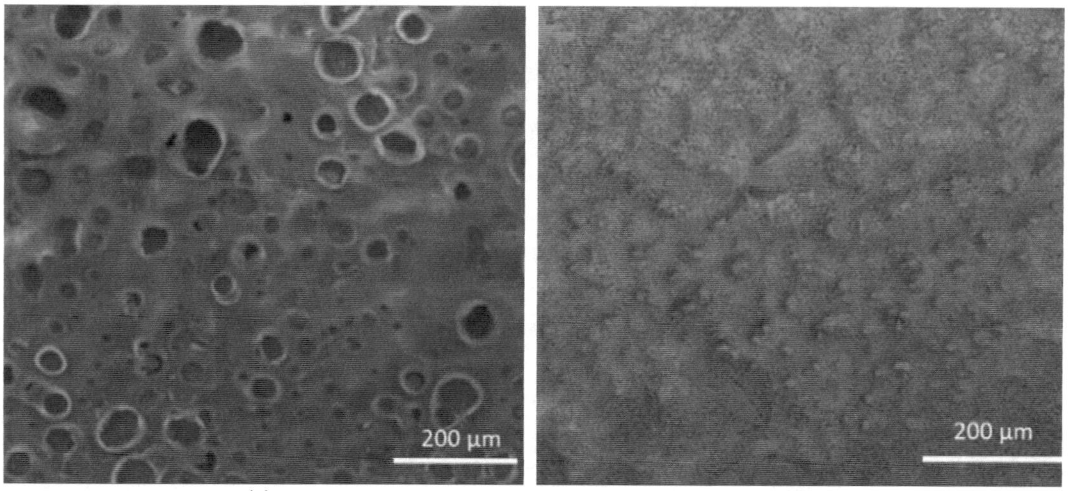

(a) (b)

Figure 6. SEM images of the samples with SiC concentration of 15 (a) and 20 (b) % v/v after high-temperature testing.

From Fig.6a it is obvious that sample has char layer with small craters after ablation. It is possible that craters were formed due to melting and liquid layer boiling during the exposure to hot gas. With that, sample Fig. 6b has flawless structure after plasma exposure.

Tensile testing showed that tensile strength of the multilayered $HfC-ZrB_2-SiC-ZrO_2$ materials during diametral compression before high-temperature testing was at 49±4 MPa and after - 68±9 MPa. Increase of tensile strength is the evidence of self-healing processes in the materials.

Analysis of the break surface (Fig. 7) reveals no delamination of the heat-protective material after rupture test before and after the exposure to hypersonic high-enthalpy oxygen-containing plasma flow for 1000 seconds.

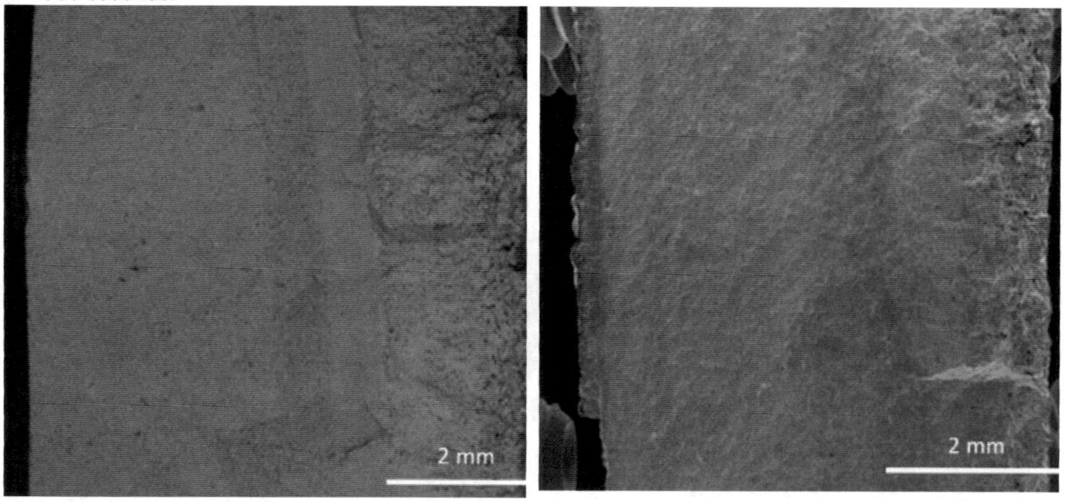

(a) (b)

Figure 7. SEM images of the fracture surfaces: a) before and b) after high-temperature testing.

4. Conclusion

Multilayered heat-protective materials with decreasing (from top to bottom side) heat conductivity were obtained. After high-temperature testing, no destruction of any HfC-ZrB_2-SiC-ZrO_2 samples was observed.

It was found, that after high-temperature testing the strength of the obtained materials during diametral compression increased for 30%. That fact leads to the conclusion that self-healing processes take place in the developed composite materials.

Conducted studies confirm the potential of the obtained multilayered HfC-ZrB_2-SiC-ZrO_2 ceramic materials as a part of heat-protection in conditions of extreme temperatures.

Acknowledgments

This work was carried out according to the Program III.23.2.3. of Fundamental Scientific Research of the State Academies of Sciences for 2013-2020 and in the framework of the Program of increasing the competitiveness of TSU.

References

[1] Eakins E, Jayaseelan D D and Lee W E 2011 Toward oxidation-resistant ZrB2-SiC ultra high temperature ceramics *Metall. Mater. Trans. A Phys. Metall. Mater. Sci.* **42** 878–87

[2] Fahrenholtz W G, Hilmas G E, Talmy I G and Zaykoski J A 2007 Refractory diborides of zirconium and hafnium *J. Am. Ceram. Soc.* **90** 1347–64

[3] Medri V, Monteverde F, Balbo A and Bellosi A 2005 Comparison of ZrB2-ZrC-SiC composites fabricated by spark plasma sintering and hot-pressing *Adv. Eng. Mater.* **7** 159–63

[4] Mashhadi M, Khaksari H and Safi S 2015 Pressureless sintering behavior and mechanical properties of ZrB2-SiC composites: Effect of SiC content and particle size *J. Mater. Res. Technol.* **4** 416–22

[5] Justin J F, Jankowiak a. and J.F.Justin A J 2011 Ultra High Temperature Ceramics: Densification, Properties and Thermal Stability *Aerosp. Lab J.* **8** 1–11

[6] Bongiorno A, Först C J, Kalia R K, Li J, Marschall J, Nakano A, Opeka M M, Talmy I G, Vashishta P and Yip S 2006 Bongiorno-ZrB2SiC-MRSB06 **31** 9

IOP Conf. Series: Materials Science and Engineering **696** (2019) 012009 doi:10.1088/1757-899X/696/1/012009

Low-Temperature Compositions with Two Gel-forming Components for Water Shut-off Purposes and Oil Recovery Increase

U V Chernova[1,2], V V Kozlov[1,2], L K Altunina[1,2], L A Stasyeva[1], V A Kuvshinov[1] and M R Sholidodov[2]

[1]Institute of Petroleum Chemistry, Siberian Branch of the Russian Academy of Sciences (IPC SB RAS) 4, Akademicheskiy Avenue, 634055, Tomsk, Russia
[2]National Research Tomsk State University, 634050 Tomsk, Lenin Avenue, 36

E-mails: chernova489@gmail.com; kozlov_vv2004@mail.ru; alk@ipc.tsc.ru; lkhn304@ipc.tsc.ru; vak2@ipc.tsc.ru; sholidodov93@inbox.ru

Abstract. This paper presents the results of laboratory studies on the developed low-temperature nanostructured gel-forming compositions for water shut-off purposes. Kinetic and rheological tests of different gel-forming compositions have been carried out. The physical model is devised, and filtration characteristics of the compositions are studied. Ultimately, the efficiency of their application to limit water inflow and to increase the oil displacement coefficient is investigated.

1. Introduction

Watercut in the Russian oil industry as a whole is steadily increasing. Extraction of associated water requires large additional costs, which leads to an increase in the cost of oil as the watering of wells increases, forcing the premature transition of steam mechanized methods of production. Operation of a large number of wells, both oil and gas, is unprofitable due to the high watering of products and due to the economic inexpediency of further operation. The works aimed at studying this issue and developing gel-forming plugging compositions are particularly significant [1-3]. On the basis of a set of laboratory studies, The Institute of Petroleum Chemistry of the Siberian Branch of the Russian Academy of Sciences (Tomsk) previously created high-temperature (60-200 °C) nanostructured gel-forming compositions of MEGA and MEGA-PRO with two gel-forming components – polymeric and inorganic – with the improved rheological characteristics. Additionally, they developed technologies of their application for water shut-off purposes and for oil recovery increase under steam-heat and steam-cyclic impact [4]. The factor causing gelation is the thermal energy of formation or of the injected heat-transfer agent, without a crosslinking agent. The MEGA and MEGA-PRO compositions are developed on the basis of the "aluminum salt – cellulose ether – urea – water" system that consists of two thermotropic gelling systems: 1) "cellulose ether – urea – water", thermotropic polymer solutions with low critical solution temperature (LCST), forming gels due to the reversible phase transition and 2) the inorganic system "aluminum salt – urea – water" forming gels due to the reaction of hydroxycondensation of aluminum ions. They form cohesively dispersed nanoscale structures of the "gel in gel" type.

The *in-situ* formed gels of the reservoir inhibit the water or steam breakthrough from the injection into the producing wells, redistribute the filtration flow of reservoir fluids in the bed, which leads to stabilization or reduction in water production of the surrounding production or CSS wells and increases oil production. The first field tests of the technology with the use of the nanostructured gelling composition MEGA for water shut-off and enhanced oil recovery were successfully carried out in 2016-2017 at 9 producing wells of the Permo-Carboniferous deposit of the Usinskoye oilfield during CSS and steam flooding techniques. The volume of the injected composition was 80-120 m^3 per well. After well treatment, a significant decrease in watering (by 12–40%) and a fold increase in oil flow rates were observed. The technology was recommended for industrial use.

MEGA-NT and MEGA-NT-PRO gel compositions have been developed to control filtration flows, to increase oil recovery and to sweep the efficiency of cooled reservoirs at 20-70 °C for low-temperature viscous oilfields. The difference between low-temperature gelling compositions and MEGA and MEGA-PRO compositions lies in the presence of a temperature regulating agent, which makes it possible to obtain aluminum hydroxide gels in formations with a lower formation temperature.

This paper presents the results of laboratory studies of MEGA-NT and MEGA-NT-PRO low-temperature gelling compositions.

2. Results and Discussion

The HAAKE Viscotester iQ rheometer (measuring system of coaxial cylinders CC25 DIN/Ti) was used as a rotary viscometry method to study gel kinetics. The solutions of MEGA-NT and MEGA-NT-PRO with different amount of thermoregulating agent (2, 3, 4 % wt.) were placed into the temperature-controlled cell of the viscometer and were (at continuous viscosity measurement) thermoregulated at different temperatures for a certain time. Viscosity measurements were made at shear rate 3 s^{-1}. The time of gel formation in solutions was determined by a sharp increase in viscosity due to the emergence of gel.

The mechanism of gel formation is different for solutions of the MEGA-NT and MEGA-NT-PRO in the temperature range above and below the LCST of the cellulose ether. When a solution of the composition is heated above the LCST of cellulose ether, in the temperature range of 60-70 ° C, a polymer gel is first formed in the system due to a solution-gel phase transition, the polymer gel first forms, and the viscosity increases sharply. Following that, inside the polymer gel, the products of carbamide hydrolysis are initiated and an aluminum hydroxide gel is formed – that is, a gel-in-gel structure, which is accompanied by a further increase in viscosity.

When a solution of MEGA-NT and MEGA-NT-PRO is thermostated at a temperature below LCST of cellulose ether, a combined aluminum hydroxide gel is formed in the temperature range of 20-40 °C and additionally strengthened by reinforcing polymer molecules of cellulose ether and soluble carbamide-formaldehyde resin, which is accompanied by a sharp increase of viscosity.

As for the temperature range of 40-60 °C, both of these mechanisms of gel formation can be implemented depending on the concentration of the components of the composition and on the temperature regulating agent. In solutions of gelling compositions containing 3-4% wt. of temperature regulating agent, a combined gel is quickly formed. The solutions containing 2% wt. of temperature regulating agent a combined gel is formed after 6-7 hours. Moreover, a polymer gel of cellulose ether is formed more quickly in solutions of MEGA-NT-PRO compositions than in MEGA-NT solutions, but more slowly a combined gel.

The rheological properties of the solutions of the gelling compositions MEGA-NT and MEGA-NT-PRO before and after thermostating at different temperatures were studied by the method of rotational viscometry at a temperature of 20 °C at various shear rates from 1 to 1200 s^{-1}(Fig. 1 and 2). The gelling compositions before gelling are pseudoplastic fluids, the dependence of stress on shear rate is non-linear, and the viscosity depends on shear rates. The gels obtained from solutions of MEGA-NT and MEGA-NT-PRO compositions are viscous-plastic and have a spatial structure capable of resisting shear stress until its value exceeds the value of the critical (limiting) shear stress. Studies of the kinetics of gelation and the rheological properties of the solutions and gels of low-temperature gelling compositions MEGA-NT and MEGA-NT-PRO with two gelling components in the temperature range of 20 – 70 °C showed that obtained gel is the synergistic bonded nanoscale structure. Compared with one-component gels, this one has increased viscosity and elasticity.

IOP Conf. Series: Materials Science and Engineering **696** (2019) 012009 doi:10.1088/1757-899X/696/1/012009

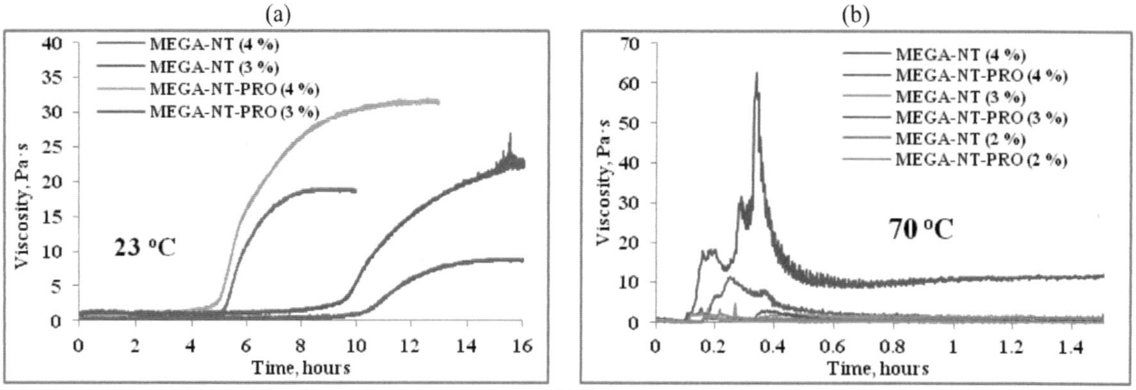

Figure 1. Gelation kinetics at 23 (a) and 70 °C (b) in solutions of the composition MEGA-NT and MEGA-NT-PRO with different concentrations of temperature regulating agent measured by rotational viscometry.

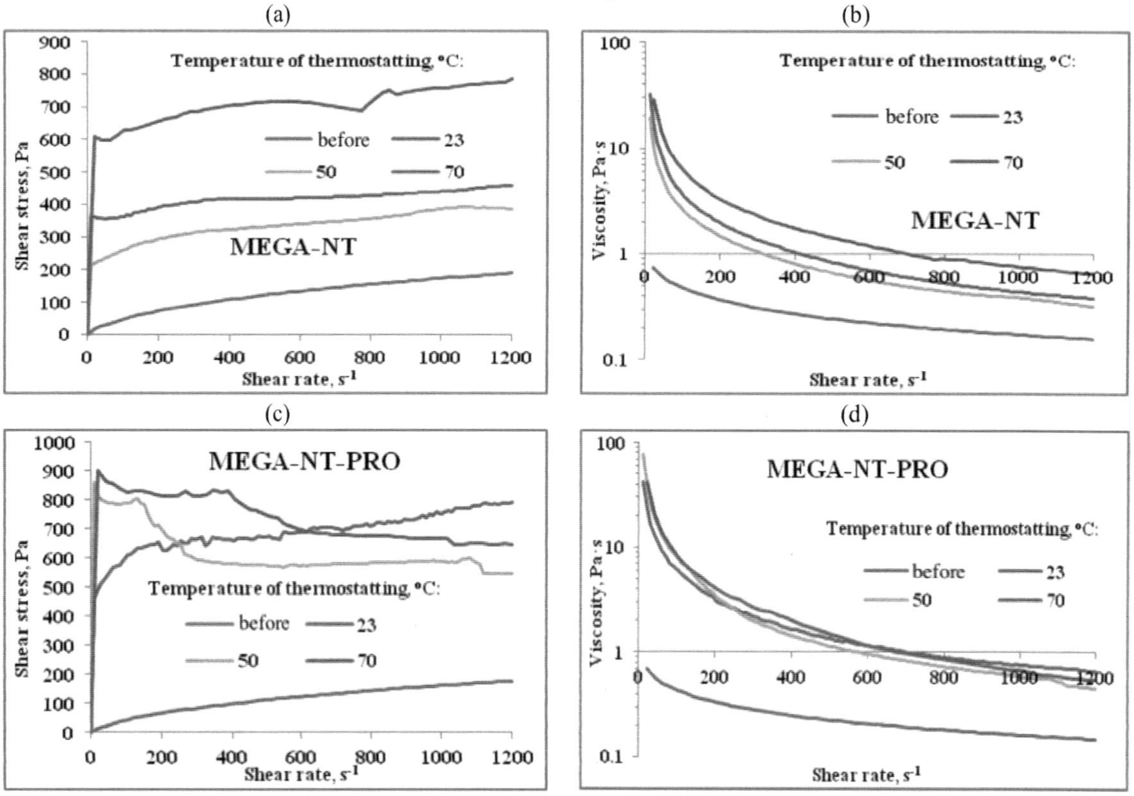

Figure 2. Rheological flow curves (a, c) and viscosity (b, d) as a function of shear rate of solutions of MEGA-NT and MEGA-NT-PRO compositions.

To assess the oil recovery ability of MEGA-NT and MEGA-NT-PRO, the experiments were carried out at the facility to study the filtration characteristics of heterogeneous reservoir models both in natural operation mode and upon pumping hot water into the Permian-Carboniferous reservoir of the Usinskoye oilfield. Four heterogeneous reservoir models were prepared, consisting of two parallel columns filled with disintegrated core material and having different permeabilities. Figure 3 and Table 1 show the results of filtration tests of the MEGA-NT composition on a heterogeneous reservoir model.

In all the experiments, the presence of gelling compositions results in the leveling of filtration flows and in the additional washout of residual oil. The increase in oil displacement coefficients was in the range of 4.6–6.56% for a higher permeability column and 5.4–40.7 for a lower permeability column.

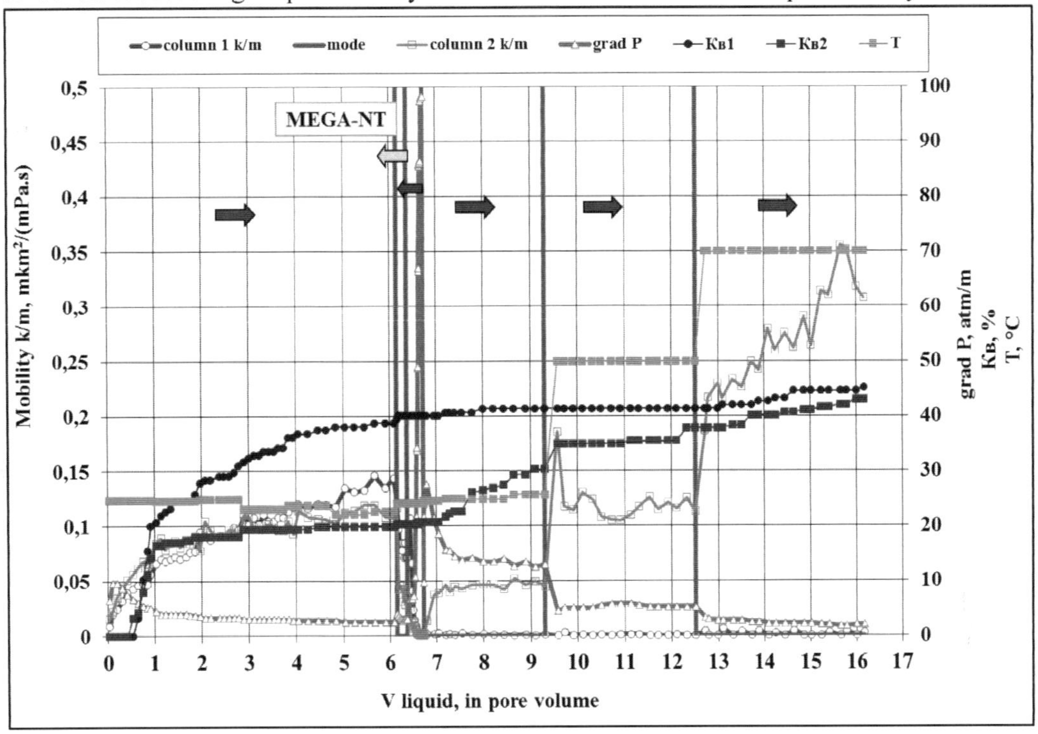

Figure 3. Influence of injection of low-temperature nanostructured gelling MEGA-NT composition on filtration characteristics and oil displacement ratio of a model of a heterogeneous Permian-Carboniferous deposit of the Usinskoye oilfield prepared from disintegrated core material with a permeability of columns of 1.608 and 0.808 μm^2.

Table 1. Effect of gelling compositions MEGA-NT and MEGA-NT-PRO on oil recovery coefficients from non-uniform reservoir models

Experience No.	Gas permeability, μm^2		Oil displacement ratio, %		The increase of oil displacement ratio,%	
	1 column	2 column	1 column	2 column	1 column	2 column
1	1.608	0.808	45.23	43.00	6.46	23.15
2	3.275	1.293	46.01	40.65	4.60	40.65
3	1.830	0.892	50.30	47.35	4.98	6.61
4	3.061	1.087	65.64	48.52	5.48	5.39

Thus, the experiments aimed at studying the filtration characteristics of a heterogeneous reservoir model (using low-temperature MEGA-NT and MEGA-NT-PRO nanostructured gelling compositions) have shown their high efficiency in blocking water breakthroughs and limiting water inflow at low temperatures, which contributes to an increase in oil recovery, which is especially significant for lower-permeability reservoir models.

3. Conclusions

MEGA-NT and MEGA-NT-PRO nanostructured gel-forming compositions with improved structural and mechanical properties have been developed. They lead to the appearance of nanostructured gels directly inside the formation. At the same time, there is a significant decrease in fluid mobility and the levelling / redistribution of filtration flows, which leads to a selective limitation of water inflow (gels can withstand pressure gradients of 60-105 atm/m), to changes in the direction of filtration flows, a to a decrease in watercut, thus limiting the breakthroughs of the injected working agent into the producing wells. The expected result is intensification of oil production and increase of oil recovery factor. The MEGA-NT and MEGA-NT-Pro low-temperature compositions have improved structural and mechanical properties and can be applied for deflecting screens in oil formations, for redistribution of filtration flows, for increasing oil recovery and limiting water inflow, as well as for waterproofing of underground workings and hydraulic structures.

Acknowledgments
The work was carried out with the financial support by the Ministry of Education and Science of the Russian Federation within the framework of the Federal Program (Project № V.46.2.3).

References
[1] Al-Muntasheri G A, Sierra L, Garzon F O and Lynn J D 2004 *Saudi Aramco Journal of Technology Summer*
[2] Sydansk R D and Seright R S 2007 *SPEPOJ* **22 (2)** 236-47
[3] Telin A G, Arslanov I R, Koptyaeva E I and Karazeev D V 2017 *Neft, Gaz, Novatsii* **8** 58-62
[4] Altunina L K, Kuvshinov V A, Stasieva L A and Kuvshinov I V 2018 *Chem. for Sustainable Dev.* **26(3)** 261-277

IOP Conf. Series: Materials Science and Engineering **696** (2019) 012010 doi:10.1088/1757-899X/696/1/012010

Obtained Heteromodulus Composites Based on Zirconium Carbide

Yu A Mirovoy[1,3], A G Burlachenko[1], S P Buyakova[1,2,3] and S N Kulkov[1,2,3]

[1]Institute of Strength Physics and Materials Science SB RAS, Tomsk, 634021, Russia
[2]National Research Tomsk State University, Tomsk, 634050, Russia
[3]National Research Tomsk Polytechnic University, 634050, Russia

E-mail: y.a.mirovoy@gmail.com.

Abstract. The article is devoted to the impact of carbon additives on densification and mechanical properties of ZrC/C heteromodulus composites. In case of addition of small amount of carbon to ZrC residual porosity decreases, whereas higher carbon content leads to more than 3vol.% increase in residual porosity. Moreover, increase in carbon content in ZrC/C composites boosts toughness of sintered materials. For example, K_{IC}=4.3 MPa×m$^{1/2}$ for ZrC and K_{IC}=7 MPa×m$^{1/2}$. for 3 vol.% C. For 15 vol.% of C, toughness is the lowest and equals 5.3 MPa×m$^{1/2}$. The extrapolation of the dependence of toughness on the non-porous state of materials is K_{IC0}=13.2 MPa×m$^{1/2}$.

1. Introduction

High melting point, oxidation resistance, chemical inertness and thermal conductivity make zirconium carbide (ZrC) a promising material for a wide range of applications. However, ZrC has a very low fracture toughness, K_{IC}≈3-4 MPa×m$^{1/2}$ [1,2] and, consequently, uncontrolled material damage accumulation significantly restricts its application field. Improvement of fracture toughness in brittle ceramics can be achieved by creation of internal dissipative structures restricting crack propagation [3]. One of the most effective mechanisms of internal dissipative structure creation and composite fracture toughness improvement is the formation of relatively weak internal interfaces between the matrix and the reinforcing component in ceramic composites. These mechanisms were described by J. Cook and J.E. Gordon [4] and their main feature is that crack tips experience branching on the matrix/particle interface under condition that particle/matrix adhesion energy is lower than that of the matrix cohesion energy. It means that the required composite structure is to be made of components with sufficiently different elastic modulus. Such materials are commonly classified as heteromodulus composite materials in which matrix and reinforcement particles possess significantly different elastic modulus [5–7]. It is suggested that ZrC-based composites may obtain a lower elastic modulus in case of technical-grade carbon addition. Zirconium carbide and carbon have E_{ZrC}=420-480 GPa [8], and E_C=2-4 GPa elastic modulus [9] respectively.

2. Experimental

The aim of this article is to study the effects of carbon addition on porosity and mechanical properties of ZrC/C composites. Commercial powders of zirconium carbide with particle mean size of 1.4 μm and standard deviation of 1.2 μm as well as technical-grade carbon <d>=1.2 μm, σ=1.5 μm were used in the research. According to X-ray diffraction analysis ZrC_x has almost stochiometric composition with X=0.98. Carbon concentration varied from 1 to 15 vol.%. Mixing of ZrC and C powders was performed in ethanol solution in the ball- milling mixer. Samples were obtained by hot pressing in argon atmosphere at 25 MPa pressure and 1800 °C temperature during 15 minutes. The composite structures were studied using X-ray diffractometer with filtered CuKα radiation and TESCAN Vega 3 scanning electron microscope at accelerating voltage of 20 or 30 kV. Hardness was measured using Berkovich pyramid nanoindentation at loading of 500 mN by NanoIndenter G200 nanoindenter according to Vickers indentation pyramid. Elastic modulus has been calculated by measuring ultrasound velocity in ceramic samples using TDS 220 test equipment.

3. Results and discussion

Addition of carbon has ambiguous impact on ZrC densification. Residual porosity of ZrC samples obtained by hot 1800 °C pressing at 23 MPa pressure during 15 minutes was 10 ± 4 %. Carbon addition in low concentration results in a better sintering behavior of ZrC and, thus, decreases the residual porosity. Porosity of ZrC/C composites with 1 and 3 vol.% carbon concentration is 5 ± 4%. On the contrary, carbon addition in higher than 3 vol. % concentration leads to increase in porosity. ZrC/C heteromodulus composites with 5 and 10 vol.% carbon concentration demonstrate slightly lower porosity than those of ZrC ceramics (8 ± 4%). For samples of ZrC/C composites with free carbon concentration of 15 vol. % porosity reaches 18 ± 4 %.

Densification of ZrC ceramics and ZrC/C heteromodulus composites demonstrates dual effect of free carbon addition. Positive influence of low carbon concentration on ZrC densification is connected with the removal of adsorbed oxygen [10]. More than 3 vol.% increase in porosity of ZrC/C heteromodulus composites with C concentration may be attributed to reduction of diffusion mass transfer due to filling of carbon vacancies. Similar discoveries concerning influence of free carbon concentration on ZrC sintering behavior were made by X.G. Wang [11].

Significant difference between ZrC ceramics' hardness and its elastic modulus calculated by various methods is discovered. For ZrC/C composites the difference is considerably lower Fig. 1. For ZrC ceramics hardness H_{Berk} is found to be 21 ± 2 GPa and Hv – 13 ± 2 GPa, while elastic modulus E_{Berk} is 420 ± 30 GPa. Elastic modulus E calculated by ultrasound transmission velocity is 220 ± 30 GPa.

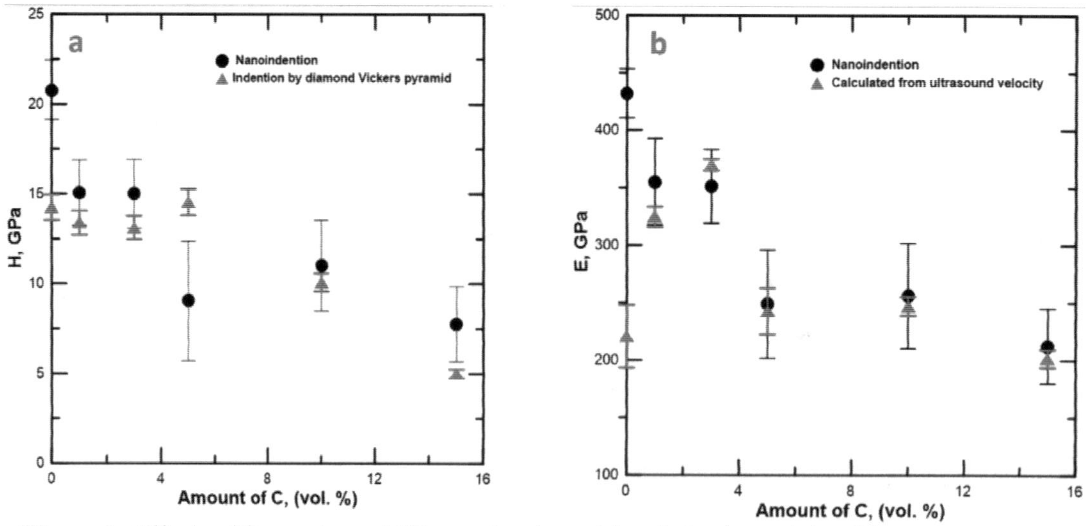

Figure 1. Effects of free carbon addition on hardness (a) and elastic modulus (b).

Needless to say, that free carbon concentration in a heteromodulus composite affects the composite's mechanical characteristics. Adding free carbon in concentrations from 1 to 15 vol.% was accompanied by hardness and elastic modulus reduction from 15 to 5 GPa and 330 to 200 GPa respectively.

According to the Vickers hardness data it is possible to calculate the stress intensity factor, K_{IC}, for sintered composites Fig. 2, using the Niihara, Morena and Hasselman formula [12]:

$$K_{Ic} = \frac{0.035\cdot(H\cdot a^{0,5})\cdot\left(E\cdot\frac{\varphi}{H}\right)^{0,4}\cdot\left(\frac{l}{a}\right)^{-0,5}}{\varphi} \tag{1}$$

where H is the hardness, E_{eff} is the modulus of elasticity, a – ½ of diagonal of impression, l – crack length, φ – constant (≈3).

Calculations conducted in accordance with the given formula allowed revealing that addition of free carbon to ZrC resulted in improved fracture toughness. For example, K_{IC} of ZrC ceramics was 4.3 ± 0.2 MPa×m$^{1/2}$, while fracture toughness of ZrC/C samples with 3 vol. % free carbon concentration was 7 ± 0,5 MPa×m$^{1/2}$ Fig. 2a. However, K_{IC} of these composites was dependent on free carbon concentration as well as other characteristics. ZrC/C composite fracture toughness with 15 vol.% carbon concentration was the lowest 5 ± 0,3 MPa×m$^{1/2}$ although this value is higher as compared to that of ZrC without carbon addition.

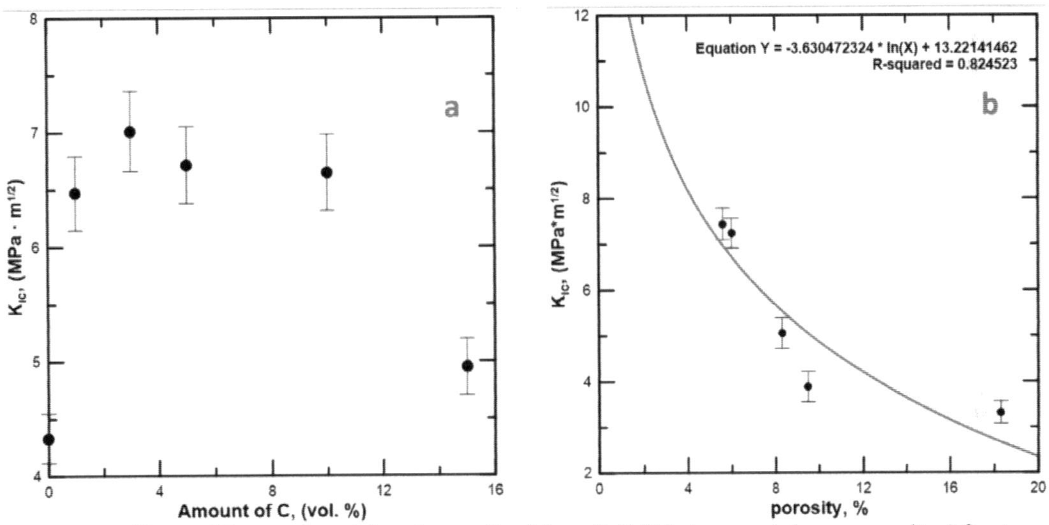

Figure 2. Effects of free carbon (a) and porosity (b) on ZrC/C heteromodulus composites' fracture toughness.

Undoubtedly, fracture toughness, as well as hardness and elastic modulus of ZrC/C heteromodulus composites were influenced by two factors – porosity and free carbon inclusions. Presence of carbon in ZrC matrix provided an increase in fracture toughness, but porosity was a factor of K_{IC} decrease. Extrapolation of K_{IC} = f (P) to the non-porous state (P = 0) showed that the largest fracture toughness of ZrC/C heteromodulus composites was obtained with 1 to 15 vol.% free carbon addition: K_{IC0} = 13.2 MPa×m$^{1/2}$, Fig. 2b. Such a high value of K_{IC} for heteromodulus composites can be provided by the Cook-Gordon implementation mechanism, when a crack interacts with ZrC-C interface. Therefore, we have replotted H, E and K_{IC} data as dependencies on porosity, as there are two factors influencing toughness, hardness and elastic modulus: carbon concentration and porosity. As it can be from these pictures after extrapolation for samples' poreless states the results are Hv_0 = 17.8 GPa, E_0 = 399 GPa and K_{IC} = 13.2

IOP Conf. Series: Materials Science and Engineering **696** (2019) 012010 doi:10.1088/1757-899X/696/1/012010

MPa×m$^{1/2}$. Values of H and E are similar to those from scientific data on dense ZrC- based ceramics. Moreover, a linear dependence between H and E typical for brittle materials Fig. 3a has been obtained.

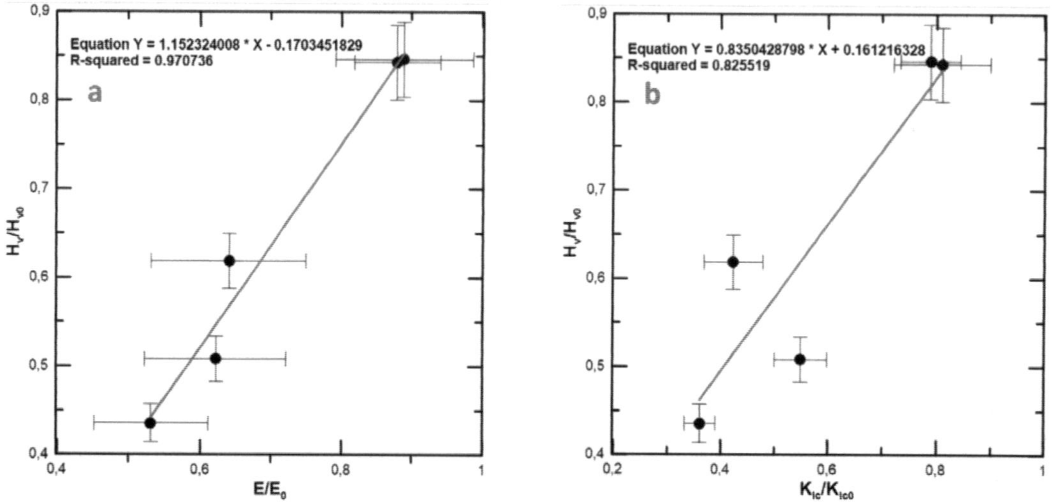

Figure 3. Correlations between Hv\Hv$_0$, E\E$_0$ (a) and Hv/Hv$_0$, K_{IC}/K_{IC0} (b) for ZrC/C heteromodulus composites

Fracture toughness for the given ZrC/C composites especially their extrapolated values is similar to those of other CMC and MMC ZrC-based composites. In some cases it exceeds K_{IC} values of ZrC-containing multi-phase ceramic composites.

For ZrC/C heteromodulus composites, a positive correlation between hardness and fracture toughness in the relative coordinates Hv/Hv$_0$ - K_{IC}/K_{IC0} has been obtained, Fig. 3b. In ZrC/C heteromodular composites hardness increase and fracture toughness of the Hv/Hv$_0$ – K_{IC}/K_{IC0} dependencies are observed with C volume concentration rise from 1 to 15%. It is suggested that correlation between hardness and fracture toughness in such composites with a wide range of free carbon concentration options might have two realizations: the increasing one with simultaneous increase in Hv and K_{IC} and decreasing one involving the decrease of Hv along with increase in K_{IC}. Most likely, increase of ZrC/C composites' hardness is connected with condition $H=H_0+K_{HP}\times L^{-1/2}$ [13], where H$_0$ is a variable corresponding to monocrystal hardness, K_{HP} is the Hall-Petch coefficient, L is the grain size. Previously, it was observed that addition of free carbon to ZrC reduces carbide grain growth [11]. Obviously, the carbide grain refinement leads to the observed increase in hardness with rising concentration of free carbon. Fracture toughness of ZrC/C heteromodulus composites increases due to the contribution of carbon fraction in accordance with energy dissipation mechanisms of crack branching at internal interfaces.

4. Conclusion

The research proves that in hot pressed ZrC/C composites an increased toughness up to 7 MPa×m1/2 can be obtained in spite of porosity up to 10% and without decreasing hardness. It shows that in such composites increased hardness and toughness can be obtained simultaneously. It is probably due to realization of both the Hall-Petch behavior when carbon addition to ceramic matrix reduces carbide grains and the Cook- Gordon mechanism.

Acknowledgments

This work was carried out according to the Program III.23 of Fundamental Scientific Research of the State Academies of Sciences for 2013-2020 and in the framework of the Program of increasing the competitiveness of TSU.

References

[1] Schönfeld K, Martin H-P and Michaelis A 2017 Pressureless sintering of ZrC with variable stoichiometry *J. Adv. Ceram.* **6** 165–75

[2] Li J, Zhang Z, Wang S, Wang Y, Dai J, Zu Y and Sha J 2016 Densification and characterization of hot-pressed ZrC-based composite doped with Nb and CNT *Mater. Des.* **104** 43–50

[3] Wei T, Fan Z, Luo G and Wei F 2008 A new structure for multi-walled carbon nanotubes reinforced alumina nanocomposite with high strength and toughness *Mater. Lett.* **62** 641–4

[4] Cook J, Gordon J E, Evans C C and Marsh D M 1964 A Mechanism for the Control of Crack Propagation in All-Brittle Systems *Proc. R. Soc. A Math. Phys. Eng. Sci.* **282** 508–20

[5] Pompidou S and Lamon J 2007 Analysis of crack deviation in ceramic matrix composites and multilayers based on the Cook and Gordon mechanism *Compos. Sci. Technol.* **67** 2052–60

[6] Shabalin I L, Wang Y, Krynkin A V, Umnova O V, Vishnyakov V M, Shabalin L I and Churkin V K 2010 Physicomechanical properties of ultrahigh temperature heteromodulus ceramics based on group 4 transition metal carbides *Adv. Appl. Ceram.* **109** 405–15

[7] Gömze L A and Gömze L N 2010 Mechanical stress relaxation in hetero-modulus, hetero-viscous complex ceramic materials *Epa. - J. Silic. Based Compos. Mater.* **62** 98–101

[8] Landwehr S E, Hilmas G E, Fahrenholtz W G, Talmy I G and DiPietro S G 2008 Microstructure and mechanical characterization of ZrC–Mo cermets produced by hot isostatic pressing *Mater. Sci. Eng. A* **497** 79–86

[9] Cost J R, Janowski K R and Rossi R C 1968 Elastic properties of isotropic graphite *Philos. Mag.* **17** 851–4

[10] Zhao L, Jia D, Duan X, Yang Z and Zhou Y 2011 Pressureless sintering of ZrC-based ceramics by enhancing powder sinterability *Int. J. Refract. Met. Hard Mater.* **29** 516–21

[11] Wang X-G, Guo W-M, Kan Y-M, Zhang G-J and Wang P-L 2011 Densification behavior and properties of hot-pressed ZrC ceramics with Zr and graphite additives *J. Eur. Ceram. Soc.* **31** 1103–11

[12] Niihara K, Morena R and Hasselman D P H 1982 Evaluation ofK Ic of brittle solids by the indentation method with low crack-to-indent ratios *J. Mater. Sci. Lett.* **1** 13–6

[13] Hall E O 1954 Variation of Hardness of Metals with Grain Size *Nature* **173** 948–9

IOP Conf. Series: Materials Science and Engineering **696** (2019) 012011 doi:10.1088/1757-899X/696/1/012011

Numerical investigation on burning rate of propane-air mixture

K M Moiseeva[1], A Yu Krainov[1] and D A Krainov[2]

[1]Tomsk State University, Tomsk, 634050 Russia
[2]Tomsk Polytechnic University, Tomsk, 634050 Russia

E-mail: Moiseeva_KM@t-sk.ru

Abstract. The paper presents a physico-mathematical model and calculation of the burning rate of a propane-air mixture. The problem formulation includes the equations of mass, momentum and energy conservation for the gas, supplemented by the equations of the oxidizer and fuel mass conservation in the gas. Godunov's scheme is used to solve the equation system of the model. The paper provides the result of the numerical simulation on the apparent and normal burning rates of the propane-air mixture. The obtained calculation results are in reasonably good agreement with the known studies of other authors.

1. Introduction

To design efficient burner devices capable of operating in extreme conditions, it is necessary to accurately predict the burning rate of the fuel. In our previous study [1] we have calculated the burning rate of a methane–air mixture. The calculation result accurately coincided with the data from [2] and [3]. According to the study [3], methane-air mixture begins to burn at a methane volume fraction of about 5.2% [3]. While as the propane-air mixture ignites at lower volume concentrations. Therefore, in some cases it is more convenient to use a propane-air mixture as a fuel.

The aim of this study is to develop a physico-mathematical model of a propane-air mixture combustion and conduct a numerical study on the dependence of the normal and apparent burning rate of the mixture on the excess fuel coefficient.

2. Mathematical model

The mathematical model is formulated under the following assumptions: there is a volume filled by a reactive mixture with a volume concentration of the combustible component a_f. The mixture is capable of exothermic chemical reaction when heated. Radiant heat transfer of the combustion products is neglected. The molecular dissociation of the combustion products at high temperature is assumed to be absent. The diffusion and thermal conductivity coefficients are temperature dependent. The gas constant is determined by the composition of the gas mixture. The rate of the chemical reaction depends on temperature according to the Arrhenius law, the reaction rate depends on the concentration of propane and oxygen, and is described by first-order kinetics for propane and first-order for oxygen. The gas motion due to the thermal expansion with increasing temperature is taken into account. Under the made assumtions the problem formulation has the form:
the continuity equation:

$$\frac{\partial \rho_g}{\partial t} + \frac{\partial \rho_g u_g}{\partial x} = 0, \qquad (1)$$

the impulse-conservation equation:

$$\frac{\partial(\rho_g u_g)}{\partial t}+\frac{\partial(\rho_g u_g^2+p_g)}{\partial x}=0, \tag{2}$$

the energy equation:

$$\frac{\partial\rho_g(\varepsilon_g+0.5u_g^2)}{\partial t}+\frac{\partial\left[\rho_g(\varepsilon_g+0.5u_g^2)+p_g\right]u_g}{\partial x}=\frac{\partial}{\partial x}\left(\lambda_g\frac{\partial T_g}{\partial x}\right)+Q_{ch}W_{ch}, \tag{3}$$

the mass balance equation of the combustible component in the gas:

$$\frac{\partial\rho_f}{\partial t}+\frac{\partial\rho_f u_g}{\partial x}=\frac{\partial}{\partial x}\left(D_g\rho_g\frac{\partial a_f}{\partial x}\right)-W_{ch}, \tag{4}$$

the oxygen mass balance equation in the gas:

$$\frac{\partial\rho_{ox}}{\partial t}+\frac{\partial\rho_{ox}u_g}{\partial x}=\frac{\partial}{\partial x}\left(D_g\rho_g\frac{\partial a_{ox}}{\partial x}\right)-\alpha_1 W_{ch}, \tag{5}$$

the perfect-gas law:

$$p_g=\rho_g R_g T_g \text{б} \tag{6}$$

the initial conditions:

$$T_g(x,0)=\begin{cases}T_0,\ 0\le x\le x_0\\ T_{gb},\ x_0<x\le\infty\end{cases},\ \rho_f(x,0)=\rho_{f,b},\ \rho_{ox}(x,0)=\rho_{ox,b}, u_g(x,0)=0,\ p_g(x,0)=p_{gb}, \tag{7}$$

the boundary conditions:

$$\frac{\partial\rho_g(0,t)}{\partial x}=\frac{\partial T_g(0,t)}{\partial x}=\frac{\partial\rho_f(0,t)}{\partial x}=\frac{\partial\rho_{ox}(0,t)}{\partial x}=0, \tag{8}$$

$$\frac{\partial\rho_g(\infty,t)}{\partial x}=\frac{\partial T_g(\infty,t)}{\partial x}=\frac{\partial\rho_f(\infty,t)}{\partial x}=\frac{\partial\rho_{ox}(\infty,t)}{\partial x}=0\,. \tag{9}$$

Notations: ρ_f and ρ_{ox} are the partial densities of fuel and oxygen in the mixture ($\rho_f=a_f\rho_g$, $\rho_{ox}=a_{ox}\rho_g$, where a_f, a_{ox} are the mass concentrations of the combustible component and oxygen in the gas); $\varepsilon_g=p_g/(\gamma-1)$ is the gas energy; $\gamma=c_p/c_v$ is the adiabatic exponent; ρ is the density, u is the velocity, t is the time, x is the coordinate; p is the pressure, T is the temperature; Q_{ch} is the thermal effect of the reaction; k_0 is the rate constant of the chemical reaction; $\alpha_1=\mu_{ox}\nu_{ox}/\mu_f\nu_f$ is the oxygen consumption coefficient in the reaction with propane; μ_{ox}, μ_f are the molar masses of oxygen and propane; ν_{ox}, ν_f are the stoichiometric reaction coefficients; Subscript: *b* is for the initial values of the state parameters, *g* - gas parameters, *ox* - oxygen, *f* - propane.

The chemical rate is determined by the right-hand sides of equations (3) – (5):

$$W_{ch}=\rho_f\rho_{ox}k_0\exp\left(-E_a/\left(R_u T_g\right)\right).$$

The connection between the mass concentration of the combustible component and the oxidizer in the gas with the volumetric content of the combustible component a_{vol} in the mixture is defined from the ratios:

$$a_{ox} = \frac{0.23(100 - a_{vol})\mu_{air}}{(100 - a_{vol})\mu_{air} + a_{vol}\mu_f}, \quad a_f = \frac{a_{vol}\mu_f}{(100 - a_{vol})\mu_{air} + a_{vol}\mu_f},$$

where a_{vol} is the volume concentration of propane in the mixture, μ_f is the propane molar mass, μ_{air} is the air molar mass.

The diffusion and thermal conductivity coefficients of the gas depend on temperature as in [1]:

$$\lambda_g = \lambda_0 \left(T_g / T_0\right)^{2/3}, \quad D_g = \lambda_g / \left(c_g \rho_g\right),$$

where λ_0 is the value of the thermal conductivity at T_g =300*K*.

The gas constant depends on the composition of the mixture and is defined as

$$R_g = R_u / \left(a_{ox}\mu_{ox} + a_f\mu_f + \left(1 - a_{ox} - a_f\right)\mu_{N2}\right),$$

where μ_{N2}, μ_{ox} are the molar masses of nitrogen and oxidizer.

The problem (1) – (6) with boundary and initial conditions (7) – (9) is solved numerically using Godunov's scheme [4]. The spatial step is equal to $\Delta h_i = 10^{-5} m$. The size of the computational domain is chosen so that during the ignition and the combustion wave formation, the shock wave arising from the energy release in the ignition zone does not reach the boundary of the domain. The time step is determined by Courant–Friedrichs–Lewy condition as

$$\Delta t = \min\left[\frac{0.2\,\Delta h_{\min}}{\max[c] + \max\left[|u_g|\right]}, \frac{\Delta h_{\min}^2}{\chi_g}\right],$$

where c is the sound speed, χ_g is the temperature conductivity of the gas. The algorithm of the numerical solution and the program code has been verified similarly to [1]. The error of the calculation is no more than 0.52%.

3. Results and discussion

The calculations have been performed at the following values of the thermophysical and formal-kinetic quantities: Q_{ch} = 42.45 *MJ/kg*, E_a =171.16 *kJ/mol*, R_u = 8.31 *J/(mol·K)*, $x_0 = 2\cdot10^{-3}$ *m*, T_0 = 1500 *K*, T_{gb} = 300 *K*, λ_0 = 0.025 *W/(m·K)*, $\mu_g = \left(\rho_{air}^0 a_{air,b} + \rho_f^0 a_{f,b}\right) / \left(\rho_{air}^0 a_{air,b}/\mu_{air} + \rho_f^0 a_{f,b}/\mu_f\right)$, $R_g = 8.31\mu_g^{-1}\, J/(kg\cdot K)$, α_1 = 3.64, $c_{pg} = 29.174\mu_g^{-1}\, J/(kg\cdot K)$, $c_{vg} = 20.864\mu_g^{-1}\, J/(kg\cdot K)$, $\mu_{air} = 29.04\cdot10^{-3}\, kg/mol$, $\mu_f = 44\cdot10^{-3}\, kg/mol$, $\rho_f^0 = 1.83 kg/m^3$, $\rho_{air}^0 = 1.2 kg/m^3$, p_{gb} = 0.1 *MPa*.

We have conducted the test numerical simulation to determine the preexponential factor k_0. We have tuned the factor k_0 to simulate the normal combustion rate of the propane-air mixture with the fuel excess coefficient ϕ = 0.85 according to the study [3]. The tuned value of the coefficient k_0 is equal to $5\cdot10^{10}$ $m^3/(kg\cdot s)$. In further calculations, the value of the fuel excess coefficient varies in the range of $\phi = 0.5 \div 1$. The aim of the calculations is to define the normal and apparent burning rate of the propane-air mixture. The apparent burning rate of the propane-air mixture is defined as the velocity of the coordinate corresponding to the partial density of the combustible component 10 times lower than the initial value. The normal burning rate is equal to the difference between the apparent burning rate and the gas velocity. The calculation results are shown in Figures 1, 2.

Figure 1 shows the comparison of the calculated results with the data from [3]. As it can be seen from the figure, the results do not exactly repeat the data from [3]. The reason may be both the assumptions of the mathematical model and the error of the experimental data [3].

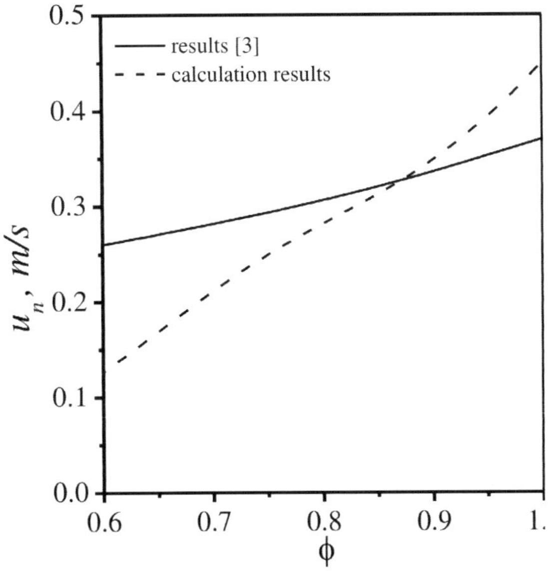

Figure 1. The normal burning rate depending on the fuel excess coefficient.

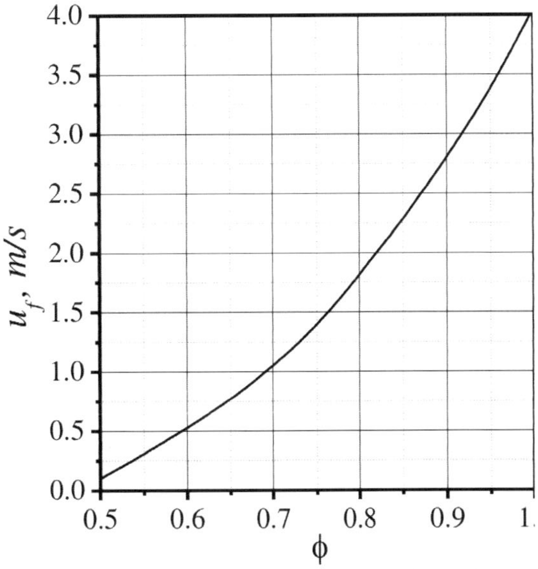

Figure 2. The apparent burning rate depending on the fuel excess coefficient.

Figure 2 shows the calculation results of the apparent flame propagation speed in the propane-air mixture. The lower flammability limit agrees with the data from [3]. For the propane-air mixture, the lower flammability limit is at the fuel excess coefficient $\phi = 0.5$, which corresponds to the 2% propane volume fraction in the mixture. According to Figure 2, the apparent burning rate of the propane-air mixture is close to zero as the fuel excess coefficient tends to the lower flammability limit.

4. Conclusions

We have developed the mathematical model of the propane-air mixture combustion. The paper provides the dependences of the visible and normal combustion front propagation velocity in the propane-air mixture on the coefficient of excess fuel in the gas. Numerical investigation has shown that for the mixtures with propane content close to the lower flammability limit, the normal and visible burning rate tends to zero. The burning rate of the poor propane-air mixtures is an order of magnitude less than the normal and visible burning rate of a stoichiometric mixture. The obtained calculation results are in reasonably good agreement with the known studies of other authors.

Acknowledgments

This work was supported by the Russian Foundation for Basic Research under Grant №19-48-703006.

References:

[1] Moiseeva K M and Krainov A Y 2018 *TSU J Math Mech* **56** 79-87
[2] Troshin K Ya, Borisov A A, Arutyunov V S, Politenkova G G and Rakhmetov A N 2013 *Russ. J Phys Chem B* **7(3)** 290-301
[3] Lewis B, Von Elbe G 1987 *Combustion, Flames and Explosions of Gases* (Moscow: Academic Press) p. 731.
[4] Godunov S K, Zabrodin A V, Ivanov M Ya, Kraiko A N and Prokopov G P 1976 *Numerical solution of multi-dimensional problems of gas dynamics* (Moscow: Nauka)

IOP Conf. Series: Materials Science and Engineering **696** (2019) 012012 doi:10.1088/1757-899X/696/1/012012

Influence of acoustic treatment and nature of solvents on viscosity of heavy oils

G I Volkova[1,2] and A V Morozova[2]
[1]Tomsk State University, 36, Lenin Avenue, Tomsk, 634050, Russia
[2]Institute of Petroleum Chemistry Russian Academy of Sciences, Siberian Branch, 4, Akademichesky Avenue, Tomsk, 634055, Russia

E-mail: galivvol@yandex.ru

Abstract. The article summarizes experimental data on the influence of acoustic treatment and the nature of solvents on the viscosity of oil from the Timan-Pechora and the Volga-Ural oil and gas-bearing basin. A combined treatment of oils, including acoustic exposure to 22 kHz and the addition of a chemical reagent was carried out. It is shown that the introduction of 1.75% wt alkaline solution of isobutyl alcohol leads to a decrease in oil viscosity of the Timan-Pechora oil and gas-bearing basin by 35%. After complex treatment (acoustic exposure for 1 min and the addition of reagent) the viscosity decreased by 60%. The introduction of alkaline chemicals into the oil of the Volga-Ural oil and gas-bearing basin does not lead to a decrease in viscosity. Oil viscosity decreases by 2.1 times after the introduction of 3% wt solvent P-12 (60% toluene, 10% xylene, 30% butyl acetate). Viscosity reduces by further 18% after acoustic exposure. Thus, the maximum viscosity reduction of heavy oils can be achieved by using the combined treatment including of acoustic exposure and addition of the selected chemical reagent.

1. Introduction

A prerequisite for the extraction of hydrocarbon resources in Siberia and the Arctic is to ensure uninterrupted operation of the transport system of the extracted volumes of oil, as well as their products [1]. Oil pipeline transport in the Russian Arctic is developed in several autonomous regions. The length of oil pipelines ranges from 63 to 523 km. For example, when developing the Novoportovskoye field on the Yamal Peninsula (one of Gazprom Neft's strategic projects) along with sea oil transportation, a pressure oil pipeline with a length of more than 100 km is used. To maintain the set temperature of the feed it was designed in the form of two insulated pipelines through which heated oil circulates between shipments.

In the Arctic regions of European Russia where there are already separate trunk oil and gas pipelines a promising way to transport hydrocarbons may be to connect peripheral wells and fields to the main pipelines. Where subsoil development projects are only being developed it is advisable to calculate the competitiveness of pipeline transport in comparison with sea and rail transportation, as well as assess its compatibility and complementarity with these types of transportation.

To date, volumes of hydrocarbon transshipment from offshore terminals in Murmansk, Vitino and Arkhangelsk are limited by the possibilities of delivering fuel to these ports via the Oktyabrskaya Railway. The prospect of the development of pipeline transport in the Russian Arctic, which will also make it possible to fully utilize the capacities of these ports and increase export volumes via the

Northern Sea Route may be connected with the connection of these ports with the onshore fields of the Russian Arctic by main pipelines.

Complications in technological processes of hydrocarbon transport, such as increasing viscosity, loss of fluidity, the formation of asphalt-resin-paraffin deposits on the surface of the equipment that occur when the temperature of the feed stream decreases, increase energy costs and reduce the efficiency of these processes. Methods used to increase the fluidity of oil traditionally include heat treatment, physical and chemical methods, dilution with light oil fractions or solvents of various nature [2-6]. Processing efficiency is often improved when using combined methods of exposure. Currently, the influence of acoustic effects in the ultrasonic wave range on the structural and mechanical properties of disperse systems is actively being investigated. Acoustic treatment technology (AT) is attracting increasing attention due to the simplicity of operation, high technology, low cost and environmental safety. However, the consequences of acoustic exposure to hydrocarbons are ambiguous and depend on the composition of the oil disperse system. So in [7] it was shown that AT of highly paraffin disperse systems leads to an increase in their viscosity and pour point. Treatment of such dispersed systems in the presence of aromatic components or polymer depressant additives eliminates the negative impact of acoustic exposure and also improves viscosity-temperature characteristics [7].

In [9], the effect of ultrasonic treatment on the viscosity and pour point of paraffinic and high-paraffin oils with different contents of tar-asphaltene components was studied. Ultrasonic action on resinous paraffin oil for 15 min allowed to lower the pour point by 16 °C and viscosity by 6 times. The relaxation of the viscosity-temperature characteristics of the oils takes several days, and the petroleum dispersed system is not completely restored, thereby partially preserving the influence of ultrasonic exposure.

The work [10] is devoted to the study of the influence of cavitation on the rheological properties of Azerbaijani oil. Based on the studies, the author claims that cavitation destroys paraffins and supramolecular structures of oil, associates, micelles and helps to reduce their size, which contributes to a decrease in viscosity. The experiments showed that as a result of cavitation processes, the dynamic viscosity of Azerbaijani oil decreased by 9% rel. However, the statement about the destruction of paraffins as a result of cavitation processes in oil is quite bold if we take into account the previously published work [11].

In practice, the rheological characteristics of high viscosity oil are usually reduced by dilution with lighter oil, gas condensate or solvents, but light hydrocarbons may be absent near the field with high viscosity oil and the introduction of organic solvents is not economically viable. Therefore, it is relevant to study complex methods of exposure including dilution of oil and AT the effectiveness of which is higher due to the presence of a synergistic effect.

In [12], the influence of ultrasonic treatment, chemical reagent treatment, and complex effects on deposits of various nature was studied. It is shown that an increase of temperature as a result of ultrasonic treatment has a positive effect on the structural and rheological parameters of the paraffin systems under study.

Thus, the above literature review shows the ambiguity of most studies, in which the fact of changes in the structural and mechanical properties of oil systems is noted, therefore, to determine the nature of the influence of acoustic effects on the behavior of oil disperse systems, additional research is required.

In this paper, we studied the effect of acoustic treatment (AT) and the nature of solvents on the viscosity of oils from the Timan-Pechora oil and gas-bearing basin (TP) and the Volga-Ural oil and gas-bearing basin (VU).

2. Experimental methods and materials

Selected oils differ in component composition and physicochemical properties (table 1).

Table 1. Viscosity, pour point and component composition of oils.

Oil	Effective viscosity at 20 °C,	T_p, °C	Content, % wt

IOP Conf. Series: Materials Science and Engineering **696** (2019) 012012 doi:10.1088/1757-899X/696/1/012012

	mPa·s		Paraffins	Resins	Asphaltenes
TP	5199	-19	59.0	31.1	9.9
VU	1014	-17	65.3	28.6	6.1

The chemical reagents used were toluene, an alkaline solution of isobutyl alcohol (IBA), Netrol acid reagent in alcohols, and R-12 solvent (60% toluene, 10% xylene, 30% butyl acetate).

AT of oil systems was performed on a ULTRASONIC TS-4M (1 kW of power) installation. 200-300 g of oil was introduced into the reactor and treated for a predetermined time.

A Brookfield DV-III ULTRA viscometer was used to study the effect of AT and chemical reagents on the rheological characteristics of the initial oils. Measurement processing was performed using Rheocalc software.

Oil viscosity measurements were carried out according to the following methods:

- AT: 300 g of the initial oil was thermostated in the cell of an ultrasonic reactor for 20 min at 20 °C. The oil was then sonicated for a predetermined time. An oil sample placed in a cell of a rotational viscometer then thermostated for 20 min and rheological measurements were performed.
- Effect of a chemical reagent: At first, the initial oil was examined. The oil sample was heated to 50 °C, thermostated for 30 min, then cooled to 20 °C, thermostated for 20 min and measured on a rotational viscometer. To assess the effect of a chemical reagent on the viscosity of oil it was dosed into hot oil and then rheological parameters were measured at various additive concentrations under the same conditions as the initial oil.
- The complex effect of AT and a chemical reagent: the viscosity of the initial oil was measured then acoustic treatment was performed and rheological parameters were measured. The chemical reagent was added and the mixture was kept for 30 min and viscosity measurements were performed. The simultaneous and sequential effect of acoustic treatment and chemicals on oil was studied. With simultaneous influence, a chemical reagent was introduced into the oil, and then acoustic treatment was carried out. With a sequential effect, the oil sample was first subjected to acoustic activation, and then a chemical reagent was added.

3. Results and Discussion

TP oil is characterized by a very high viscosity, the values of which increase significantly with decreasing temperature. Studies have shown that the rheological properties of oil change little after AT: after processing for 1-8 min, a decrease in viscosity of the original oil by 8-10% is observed.

In the presence of 3% wt toluene depression viscosity is 30%. More effectively, the viscosity of TP oil decreases after complex exposure: dilution of 3% toluene and AT for 5 min leads to a decrease in viscosity by 2.1 times (table 2).

Table 2. Change in the viscosity of VU oil depending on the method of exposure (20 °C).

Sample	Viscosity, mPa·s
TP	5200
TP + 1 min AT	5020
TP + 3% toluene	3640
TP + 3% toluene + 2 min AT	3640
TP + 3% toluene + 3 min AT	3500
TP + 3% toluene + 4 min AT	3000
TP + 3% toluene + 5 min AT	2476
TP + 3% toluene + 6 min AT	2800
TP + 3% toluene + 7 min AT	2800
TP + 3% toluene + 8 min AT	2950
TP + 3% toluene + 8 min AT_2 days[a]	2950
TP + 3% toluene + 8 min AT_3 days[a]	3150

[a] Relaxation time

IBA was added to TP oil in an amount of 0.5-5% wt. The introduction of 1.75% wt IBA reduces the viscosity of TP oil by 35%. The maximum depressant effect of 60% was achieved after the introduction of 1.75% wt IBA and subsequent treatment for 1 min (figure 1). The increase in the time of ultrasonic treatment of the mixture oil + 1.75% wt IBA up to 2 min leads to an increase in viscosity. The processed sample, aged for 3 days, relaxes, but does not reach the initial viscosity values

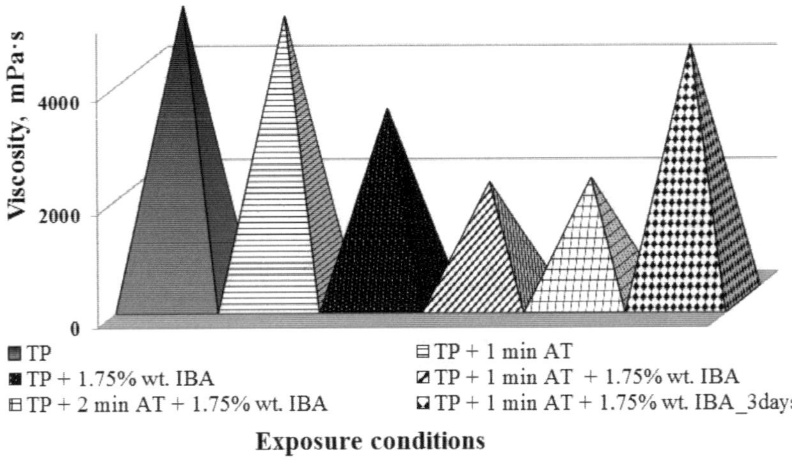

Figure 1. Change in effective viscosity (shear rate 20 s^{-1}) of TP oil after AT and dilution with IBA; temperature 20 °C.

VU oil is also characterized by a high content of resins (28.6% wt), asphaltenes (6.1% wt), and a lower acid content (concentration of COOH groups is 0.05%) compared to TP oil. This oil after AT for 2 min shows a decrease in dynamic viscosity by an average of 10% over the entire range of shear rates compared to the viscosity of the original oil. Increasing the processing time to 5 min leads to an increase in viscosity values.

The low content of petroleum acids in this oil precludes its treatment with alkaline solutions to reduce viscosity. To reduce the viscosity of VU oil Netrol acid reagent solutions in methyl, isopropyl, and isobutyl alcohols were used. It turned out that the use of an acid reagent in methanol leads to a decrease in viscosity, however the oil mixture exfoliates over time.

As can be seen from table 3, the introduction of 1 and 2% wt Netrol's solution in isopropyl alcohol (IPA) (Netrol's concentration in alcohol of 1% wt) in oil reduces the viscosity by 1.7 and 1.9 times, respectively. The increase in the content of Netrol to 6% wt leads to an increase in dynamic viscosity compared with the initial sample. AT of mixtures of oil + 1% wt and oil + 2% wt solution of Netrol in isopropyl alcohol leads to an additional decrease in viscosity by 10 and 14%, respectively (table 3).

Table 3. Change in the viscosity of VU oil depending on the method of exposure (20 °C).

Sample	Viscosity, mPa·s	
	No AT	AT, 2 min
VU	1121	1009
VU + 1% methanol (1% Netrol)	703	621
VU + 2% methanol (1% Netrol)	695	598[a]
VU +0.5% IPA (1% Netrol)	790	694
VU +1% IPA (1% Netrol)	672	610
VU +2% IPA (1% Netrol)	595	524

VU +0.5% IBA (1% Netrol)	749	622
VU +1% IBA (1% Netrol)	652	628
VU +2% IBA (1% Netrol)	557	520
VU +1% P-12	701	671
VU +2% P-12	559	490
VU +3% P-12	536	441

[a] Mixture exfoliates

The effectiveness of ultrasonic treatment increases with the introduction of R-12 solvent into oil (60% toluene, 10% xylene, 30% butyl acetate). For example, dilution of oil 3% wt solvent P-12 leads to a decrease in viscosity by 2.1 times and after acoustic exposure the viscosity decreases by another 18%.

4. Conclusion

The effect of acoustic treatment and nature of solvents on the viscosity of heavy oils has been studied. The addition of chemical reagents into the oils reduces the viscosity. Solvent effectiveness is strongly dependent on the composition of the treated substance. The combined effect of acoustic treatment and solvent results in an additional decrease in viscosity characteristics.

Thus, the maximum decrease in the viscosity of heavy oils can be achieved by using the combined effects of acoustic treatment and a selected chemical reagent.

Acknowledgments

This research was supported by the Ministry of Science and Higher Education of the Russian Federation (Project N° V.46.2.1)

References

[1] Yashchenko I G 2015 *Ecol. Bull. Russ.* **8** 32
[2] Loskutova Yu V, Prozorova I V, Yudina N V 2011 *Chem. Technol. Fuels Oils.* **47(5)** 358
[3] Musina N S, Maryutina T A 2016 *J. Anal. Chem.* **71(1)** 27
[4] Litvinetc I.V., Nebogina N A, Prozorova I V 2018 *Bull. Perm National Res. Polytech. University. Chem. Technol. Biotechnol.* **3** 42
[5] Abramov V O, Abramova A V, Bayazitov V M, Altunina L K, Gerasin A S, Pashin D M, Mason T J 2015 *Ultrason. Sonochem.* **25** 76
[6] Shakirova L N, Sudykin S N, Abdrakhmanova L M, Sudykin A N, Yakubov M R 2015 *Collection of scientific papers TatNIPIneft* (PJSC "Tatneft", Moscow) 263
[7] Anufriev R V, Volkova G I 2016 *Key Eng. Mater.* **670** 55
[8] Anufriev R V, Volkova G I, Vasilyeva A A, Petukhova A V, Usheva N V 2015 *Procedia Chemistry.* **15** 2
[9] Volkova G I, Prozorova I V, Anufriev R V, Yudina N V, Mullakaev M S, Abramov V O 2012 Oil Refining and Petrochemicals. Sci. Technol. Achievements Best Practices. **2** 3
[10] Nurulaev V Kh 2016 *Oil&Gas J. Rus.* **9** 88
[11] Yakovlev V A, Zavaruhin S G, Kuzavov V T, Stebnovskiy S I, Malih N V, Maltsev L I, Parmon V N 2010 Chem. Phy. J. **3** 43
[12] Wang Zh, Zengb J, Songc H 2017 Procedia Chemistry. **36** 162

The influence of stress states on the martensitic transformation in Y-TZP ceramic

Sablina T.Yu.[1,2], Sevostyanova I.N.[2], Bayanov A.N.[1] and Kulkov S.N.[1,2]

[1]Tomsk State University, Tomsk, Russia
[2]Institute of Strength Physics and Materials Science SB RAS, Tomsk, Russia

E-mail: Sabltat@mail.ru

Abstract. It has been studied the influence of the type of stress state arising during different types of loading - for axial compression, diametral compression (Brazilian test) and three-point bending of Y-TZP ceramics with different grain sizes. It was shown that with increasing grain size of ceramics there is an increase in strength properties. The maximum values of all strength characteristics are fixed for ceramics with a grain size of 1.55 microns. It was found that on the fracture surfaces the formation of a monoclinic phase caused by the tetragonal–monoclinic transformation under the influence of applied stresses after axial compression test, Brazilian test and three-point bending test. As the average grain size increases, the content of the monoclinic phase increases after all types of mechanical tests. For ceramics with the average grain size of 0.53 μm, regardless of the type of loading, the content of the monoclinic phase on the fracture surface was approximately 5-7%. For ceramics with a maximum grain size of 1.55 microns after axial and diametrical compression tests, the content of the transformation phase increased to 27 % and after bending tests up to 39 %. The increased content of the monoclinic phase on the fracture surface after bending tests, in comparison with the axial and diametrical compression tests, is due to a complex stress-strain state, when during deformation areas with tensile and compressive stresses are formed on the fracture surface.

1. Introduction

Ceramics based on zirconia has a special place in various fields of engineering and technology, due to its high physical, chemical and mechanical properties [1, 2]. This is especially important for devices working in Arctic region as in these conditions ceramics are exposed to various kinds of external influences, which determine their stress-strain state. It is known [3] that martensitic transformation of tetragonal (T) phase into monoclinic (M) phase in zirconia ceramics occurs under the action of applied stresses. This T-M transformation is influenced by the content of the dissolved stabilizer, the internal stresses [4] stabilizing the tetragonal phase and the grain size [5-7]. The ratio of these three factors affects the implementation of transformation toughening materials and its mechanical properties.
The aim of this paper was to study the influence of the type of stress-strain state on the martensitic transformation in zirconia ceramics with different grain sizes.

2. Materials and methods

Ceramic samples were obtained by cold uniaxial pressing under 100 MPa pressure of plasticized powder ZrO_2-5.5 wt.%Y_2O_3 with following sintering of the green body in a high temperature furnace in air. The samples were sintered at temperatures of 1500 and 1600°C during isothermal holding times

1, 3, and 5 h to obtain a ceramics with different grains size. The heating and cooling rates were 240°C/h. The sample structure was studied on a TESCAN Vega scanning electron microscope (SEM). Grain size was determined from images of thermally etched polished surfaces using ImageJ[R] program. The samples density was determined by Archimedes method for five samples with relative error not more 5%. The porosity of ceramic samples was calculated from the ratio of the measured density to the theoretical one and compared with the data obtained from SEM images of the structure. The porosity of sintered ceramics did not exceed 4%. The phase composition of samples was determined from diffraction patterns obtained on an X-ray diffractometer with filtered CuKα radiation in the range of angles 2θ = 20°–80° with a step size of 0.05° and with statistical accuracy better 3%. The mechanical tests were carried out by universal testing machine "Instron-1185" at a loading rate of 0.1 mm/min with automatic recording of the loading curve in the coordinates "load-displacement", which are then rebuilt in the deformation curves «σ-ε».
Cylindrical samples with diameter ∅=8 mm and height h=12 mm were used for axial compression tests. Samples with 5x5x35 (mm) size were used for three-point bending tests and cylindrical samples with dimensions ∅=28 mm and h=12 mm were used for diametral compression tests (Brazilian test).

3. Results and discussion

Figure 1,a shows the SEM image of the typical grain structure of zirconia ceramics sintered for 5 hours at 1600°C. The insert shows the grain sizes distribution. Histograms have a normal distribution with one maximum. With increasing of temperature and holding time during sintering, the maximum on the histograms shifts towards larger grain sizes. Figure 1,b shows the dependence of the average grain size materials sintered at 1500 and 1600°C with increasing of isothermal holding time.

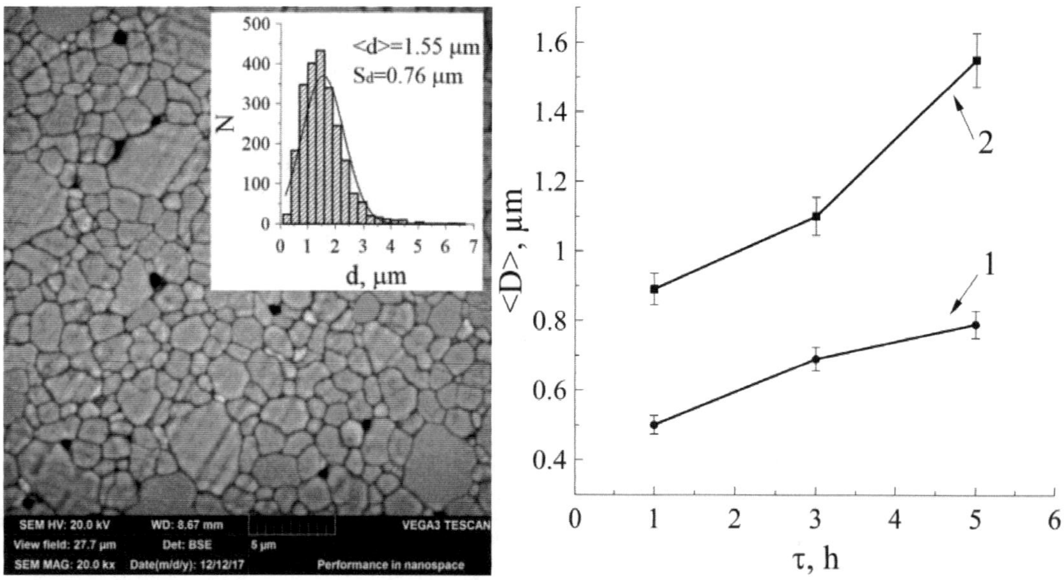

(a) (b)

Figure 1. SEM image of the typical grain structure of ceramics ZrO_2 (Y) after sintering at T=1600 °C, τ=5 h (insert shows grain size distribution) (a) and the dependence of the average grain size of ceramics sintered at a temperature of 1500°C (curve 1) and 1600°C (curve 2) vs. the isothermal holding time (b)

As one can see the average grain size increased both with an increase of the sintering temperature and with an increase of the holding time: the average grain size varied from 0.53 to 0.8 while time changes from 1 to 5 hours μm for 1500°C and from 0.9 to 1.55 μm for 1600°C with the same sintering time. According to X-ray phase analysis after sintering all samples were in two-phase state - cubic and

IOP Conf. Series: Materials Science and Engineering **696** (2019) 012013 doi:10.1088/1757-899X/696/1/012013

tetragonal phases. The content of the cubic phase in all samples approximately 20%, which agrees with the ZrO_2-Y_2O_3 phase diagram [8] and did not depend on the grain size. Despite the fact that the average grain size obtained after sintering at 1600°C for 5 hours was 1.55 μm and in accordance with [6] exceeds the critical grain size (~1.4 μm) the spontaneous transformation of the tetragonal phase to the monoclinic after cooling does not observed.

Figure 2a shows dependences of mechanical properties of sintered ceramics as a function of grain size. As one can see these dependences are similar and on figure 2b are plotted all data in relative coordinate which confirm that all mechanical properties depends on grain size of sintered material and can be described by one dependence. That is, for different stress-strain state formed in the process of compression, bending or tensile the strength of the material increases with increasing grain size.

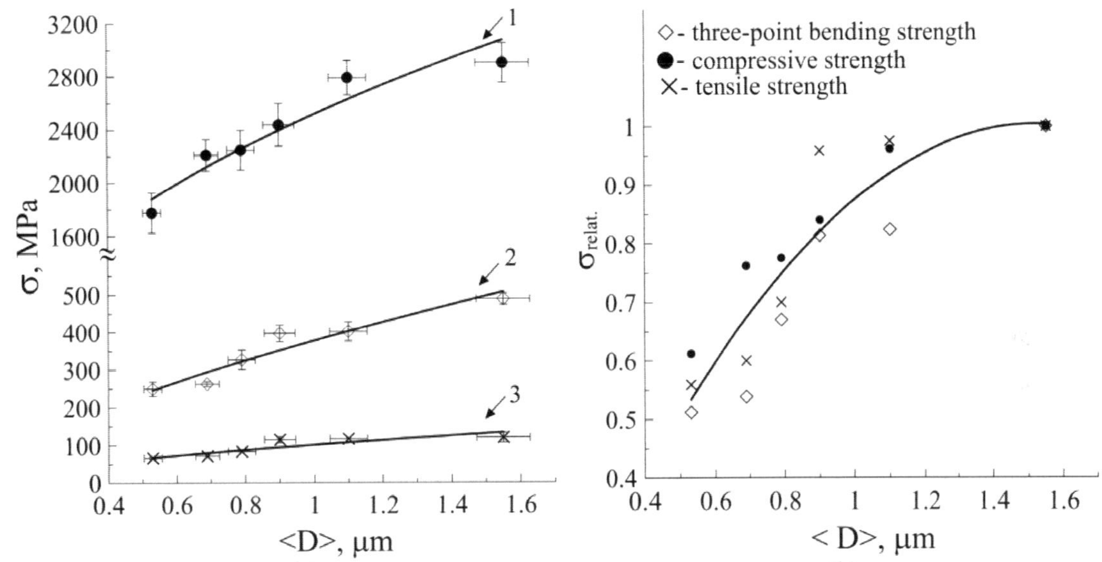

(a) (b)

Figure 2. The compression strength (curve 1), bending strength (curve 2) and tensile strength (curve 3) as a function of grain size (a) and relative strength characteristics as a function of grain size (b)

X-ray phase analysis carried out after all types of mechanical tests are shown that a monoclinic phase was formed on the fracture surface of the samples and its content increased with the increase in the average grain size (figure 3).

The dependences of content of the monoclinic phase on the fracture surfaces are similar for all types of loading but for a small grains of monoclinic phase content approximately the same and for a larger grain size changes from 27% for axial compression and diametral compression tests up to 39 % for bending test.

Figure 4 shows the dependences of the strength vs. content of the monoclinic phase on the fracture surface. These dependences are well agreed with literature data [9-13] and probably due to a transformational toughening observed in this system.

From X-ray diffraction analysis, the values of microstresses of the tetragonal phase $ZrO_2(Y)$ on the fracture surface of the samples were calculated, figure 5. As one can expect microstresses decrease with increase of grain size and the content of monoclinic phase, for a both cases an increase of grain size of tetragonal phase means cleaning grains from different defect types.

Figure 5c shows the correlation between macrostrength and microstresses, which is in a good agreement with the literature data for other oxide systems [14] and shows a high contribution of grain boundaries to the mechanical properties of ceramic.

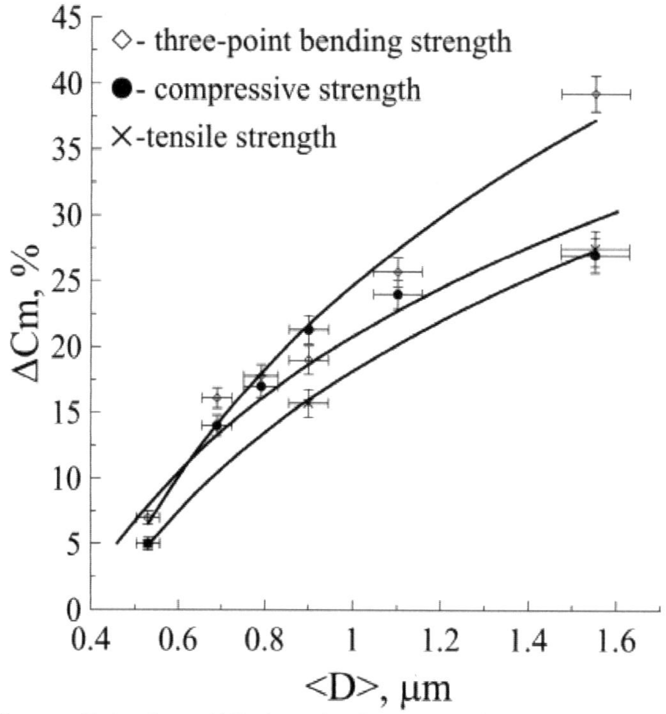

Figure 3. Content of monoclinic phase (ΔCm) on the fracture surface after different type of loading as a function of grain size

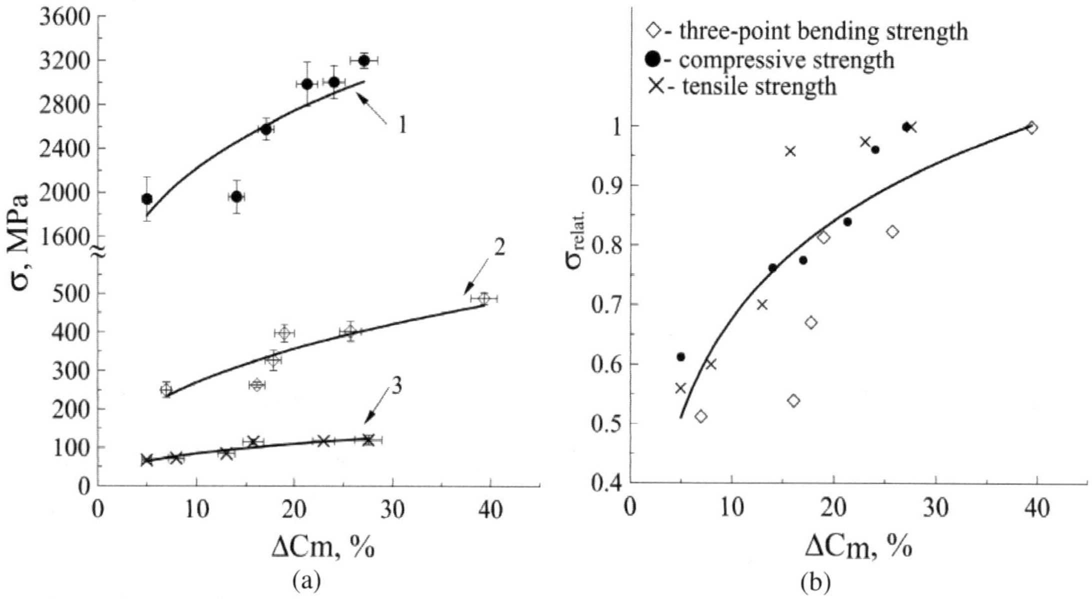

Figure 4. Dependence of compressive strength (curve 1), bending strength (curve 2) and tensile strength (curve 3) vs. the content of monoclinic phase on the fracture surface (ΔCm) (a) and dependence of relative strength characteristics vs. ΔCm (b)

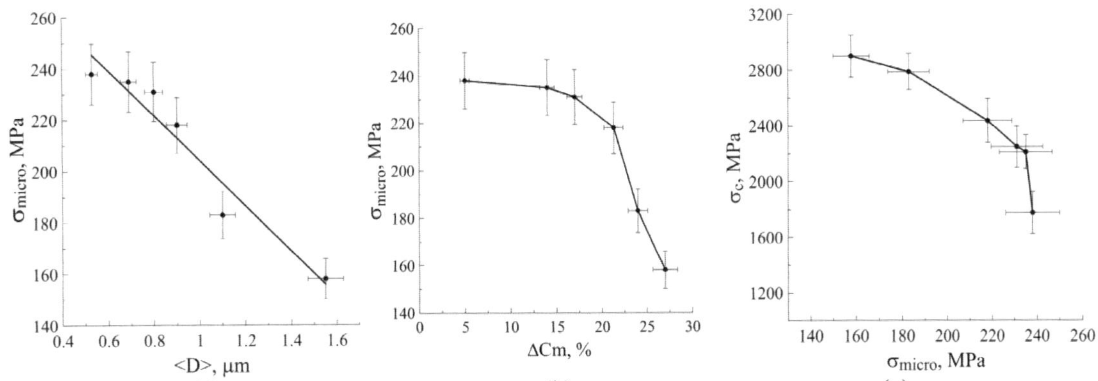

(a) (b) (c)

Figure 5. Dependence of the microstresses as a function of average grain size (a) and vs. content of the monoclinic phase on the fracture surface (b) and the dependence of the macrostrength vs. microstresses (c)

4. Conclusion

It has been shown that grain size of sintered ceramics are stipulate of its strength for different types of loading – bending test, axial and diametral compression tests and these dependences are described one function. The maximum values of all strength characteristics are fixed for ceramics with a grain size of 1.55 microns. It was found that content of monoclinic phase on the fracture surfaces has a similar dependences for all types of loading. With a larger grain size, there was a higher difference in the content of the monoclinic phase on the fracture surface for different types of loading. It has been shown that microstresses decreased with increase of grain size and has high influence on ceramic macrostrength.

Acknowledgements
This work was carried out according to the Program III.23 of Fundamental Scientific Research of the State Academies of Sciences for 2013-2020 and in the framework of the Program of increasing the competitiveness of TSU.

References

[1] Evans A.G, Cannon R.M.1986 *Acta Met.* **34**, 761–800
[2] Porter D.L, Evans A.G, HeuerA.H.1979 *Acta Met.***27,** 1649–1654
[3] Hannink, R.H.J., Kelly, P.M. Muddle, B.C. 2000 *J. Am. Cer. Soc.***83**(3), 461-487
[4] Kulkov S.N., Buyakova S.P., Gömze L.A. 2017 *J. Phys.Conf. Ser.***790** 012015
[5] Wang J, Rainforth M, Stevens R. 1989 *Br Ceram Trans J.***88**, 1–6
[6] Eichler J., Rödel J., Eisele U, et al. 2007 *J Am Ceram Soc* .**90**, 2830–2836
[7] Savchenko N.L., Sablina T.Y., Sevost'yanova I.N. et al. 2018 *Tech. Phys. Lett.* **44**, 663
[8] Duwez, P., Brown, F. H., & Odell, F. 1951 *J. of the Electrochemical Soc.* **98**(9), 356
[9] Korobenkov M.V., Kulkov S.N., Leitsin V.N., Tovpinets A.O. 2018 *Matec Web of Conf.* **243**, 00025
[10] Buyakov A.S., Vasilyeva E.O., Kulkov S.N. 2018 *Solid state physics* collection of materials of the XVI Russian scientific student conference, ed. I. Y. Litovchenko (Tomsk: NTL Publishing house) pp. 21-23
[11] Kozulin A. A., Naricowich A. S., Aliev V. D., Leitsin V.N., Kulkov S. N. 2016. Izv. Vyssh. Uchebn. Zaved. Fizika **59** (7), 108-112
[12] Shutilova E.S., Buyakova S.P., Kulkov S.N. 2015 *AIP Conference Proceedings* **1683** 020212
[13] Kanaki A.V., Buyakova S.P., Kulkov S.N. 2016 *Inorganic Materials: Applied Research.* **7** (5), 750-755
[14] Kumar S., Kurtz S. K., Agarwala V. K. 1996 *Acta mech.* **114** (1), 203-216

IOP Conf. Series: Materials Science and Engineering **696** (2019) 012014 doi:10.1088/1757-899X/696/1/012014

Structural and Mechanical Properties of Water Oil Emulsions of Highly Viscous Oils

E A Savrasova[1], N V Yudina[1], N A Nebogina[1] and G I Volkova[1,2]

1 Institute of Petroleum Chemistry, Siberian Branch, Russian Academy of Sciences, 4, Akademichesky Ave., 634055, Tomsk, Russia

2 National Research Tomsk State University, 36, Lenin Ave. 634050, Tomsk, Russia

E-mail: natal@ipc.tsc.ru

Abstract. The results of the study of the process of formation of emulsions in high viscosity oils are presented. The effect of the oil composition and the content of the aqueous phase on the microstructure, rheological properties, and activation energy of the viscous flow of oil-water emulsions are shown. Based on the analysis of the group composition of oils and the intermediate layer formed during the separation of emulsions, an increase in the mass fraction of benzene, alcohol-benzene resins, and asphaltenes in the intermediate layer is established.

1. Introduction

High-viscosity oils form emulsions that are difficult to break. The stability of oil-water emulsions is determined by the amount and composition of surfactants called emulsifiers [1, 2]. These include asphaltenes, resins, salts of naphthenic acids and heavy metals, mechanical impurities, paraffin microcrystals. These substances, adsorbed at the phase boundary, form adsorption membranes with high structural viscosity on the droplets of the dispersed phase. The adsorption sheathes prevent coalescence [3-5]. Significant differences in the quantitative and qualitative composition of the main emulsifying components in oils are reflected in the process of emulsion formation, its stability and structural and mechanical properties [6, 7].

It is known that physicochemical and structural-rheological properties of disperse oil systems are determined by the structure, size and composition of complex structural units resulted from the association of resin-asphaltene components [8-11]. To date, the nature of the processes of structure formation and their relationship with the rheological properties of oils have not yet been fully elucidated, and this explains the lack of sufficient clarity regarding the regulation of the rheological properties of high-viscosity oils and water-oil emulsions under conditions of production, transportation, and storage.

The separation of oil-water emulsions of heavy oils forming highly stable intermediate layers is one of the urgent problems in the preparation of high-viscosity oils. The introduction of new technologies for the breakdown of stable oil-water emulsions requires the development of a science-based approach based on in-depth knowledge of the formation of the structure of the intermediate layer depending on the content and composition of the oil phase.

The purpose of the work was to study the structural and mechanical properties of water-oil emulsions of 3 high-viscosity oil wells of the Usinskoye oilfield, depending on the water content and natural oil emulsifiers.

2. Experimental section

Model water/oil emulsions were prepared using a PE-00118 150-W agitator at a blade rotation speed of 1500 rpm for 10 min. As the dispersed phase for the emulsions, produced water (pH 6.8-7.0, mineralization 86.5 mg/L) was used.

The separation of the intermediate layer (i/l) from the water and oil phases was carried out using a separatory funnel. The intermediate layer was centrifuged to isolate the organic component.

Isolation of asphaltenes was carried out by the 'cold' Golde method. The content of saturated (SHs) and aromatic hydrocarbons (AHs), benzene (BR) and alcohol-benzene (ABR) resins in oil and in the intermediate layer was determined by column liquid adsorption chromatography. The mass fraction of paraffin hydrocarbons was determined by the method of urea complexation.

The dispersion of oil-water emulsions was evaluated by optical microscopy. Microphotographs of the samples were taken using an Axio Lab A1 biological microscope (Carl Zeiss) equipped with an Axiocam ERc 5s digital camera.

The rheological parameters of initial oils and emulsions were analyzed using a Brookfield LVDV-III Ultra rotational viscometer. The Arrhenius-Frenkel-Eyring equation was used to determine the activation energy of a viscous flow

$$\eta = A \cdot exp^{E/RT},$$

where η is the effective viscosity; A is the preexponential coefficient, including in a latent form the dependence of viscosity on other parameters of the structure, in particular molecular weight; E is the activation energy of the viscous flow process, representing the energy barrier of the fluctuation transition process; R is the universal gas constant; and T is the absolute temperature. The activation energy of the viscous flow Eact was determined as the angular coefficient of the linear dependence of the effective viscosity on temperature in the coordinates $\ln(\eta) = f(T^{-1})$, and that for nonlinear dependences was determined by the slope of the selected linear sections or tangent to the curve.

3. Results and Discussion

The oil from the Usinskoye oilfield is heavy oil. In Table 1 are presented the physico-chemical characteristics of the oil sampled from different wells of the Usinskoye oilfield. The highest viscosity and density is exhibited by the oil from well No. 3

Table 1. Physical and chemical properties of oil sampled from wells of the Usinskoye oilfield

Sample from the well No.	Density, kg/m^3	Viscosity, mPa·s	T_{fr}, °C
	20 °C		
1	959	4800	- 17.0
2	932	5850	- 20.0
3	972	8840	- 14.5

Emulsions formed on the basis of oil from the Usinskoye oilfield are characterized by high stability. Hence, they do not broken for a long time after formation. The main factors determining the stability of oil-water emulsions are dispersion, viscosity, electrical properties, concentration, and type of natural emulsifiers [1].

In figure 1 there are microphotographs of 10% and 50% emulsions of oil sample from the well No. 1. The average sizes of water globules (d_{av}) and dispersion (D) of all emulsions under study were calculated from microphotographs

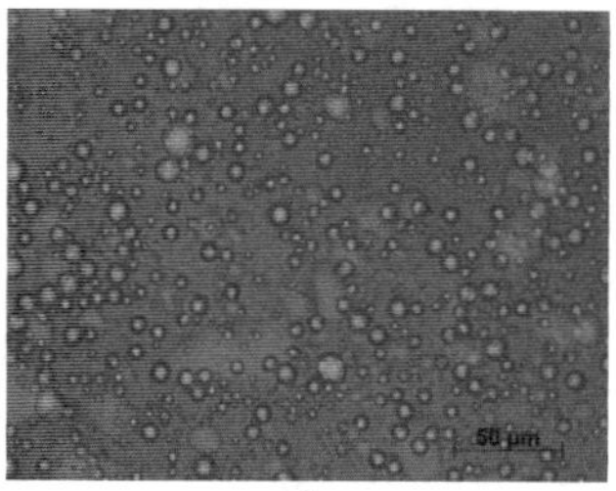

a)

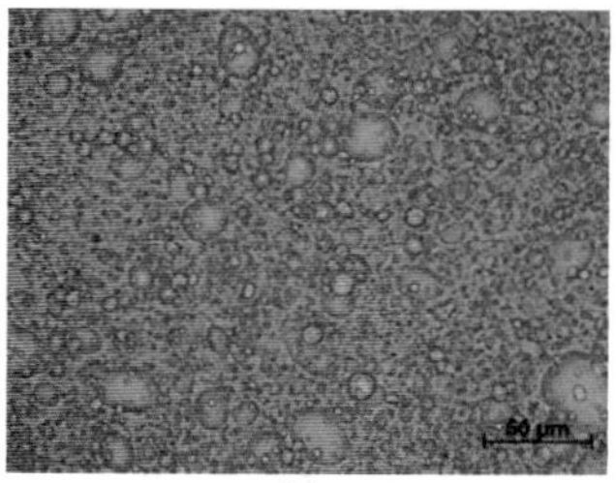

b)

Figure 1. Microphotographs of 10% (a) and 50% (b) emulsions of the oil from the well No. 1

In Table 2 we gave the average sizes of water globules in emulsions with different water contents. Globules of the lowest diameter and a higher degree of dispersion are formed in the 10% emulsion of oil sampled from the well No. 3. With an increase in the percentage of the aqueous phase up to 50%, an increase in the average droplet diameter by 2.5 times and, accordingly, a decrease in dispersion (D) are observed in the composition of emulsion.

Table 2. Average diameter of water droplets and dispersability of water- oil emulsions

Sample from the well No.	10% emulsion		50 % emulsion	
	D_{av}, µm	D, μm^{-1}	D_{av}, µm	D, μm^{-1}
1	6.5	0.15	16	0.06
2	5.0	0.20	13	0.08
3	2.0	0.50	11	0.09

The features of the formation of supramolecular structures in oils and the strength of bonds in associative complexes were evaluated from viscosity-temperature dependences. It is shown in figure 2 that the viscosity of oil sampled from the well No. 3, which is characterized by the maximum viscosity values among the objects under study and the viscosity of 10% emulsion decreases monotonically in the temperature range from +20 to +50 °C. This is determined by the size of associative structural units. With a decrease in temperature to + 10 °C, the viscosity of the objects under study increased by 3 times.

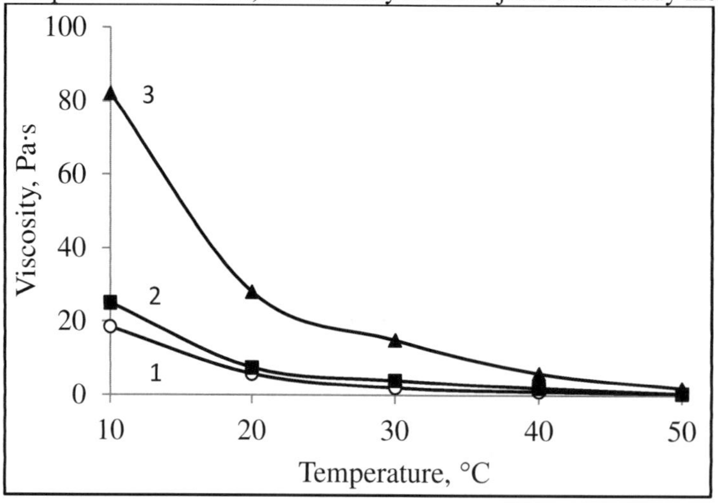

Figure 2. Temperature-viscosity curves for oil from the well No. 3 and emulsions at a shear rate of 9 s^{-1}: 1 – oil; 2 – 10% emulsion; 3 – 50% emulsion

In the case of a 50% content of the aqueous phase, the viscosity of the emulsions increases 4-5 times within the range of shear rates and temperatures compared to dry crude oil. The heating of oil and emulsions to 40-50 °C has a significant effect on viscosity, the values of which in this temperature range are reduced and become little dependent on shear load.

Based on the results of rheological measurements, the energies of activation of the viscous flow (Eact) were calculated. In Table 3 are given the values of effective viscosity at a shear rate 9 s^{-1} and the activation energy of a viscous flow for oils and emulsions. The increase in the viscosity of emulsions formed by the oil from the well No. 3 is the most critical. The droplets of emulsions formed on the basis of this oil are of a smaller diameter in comparison with the emulsions based on oil from other wells.

The graphical dependence of ln(η) on 1/T in the temperature range from +10 to ...+ 50 °C for high resin oils and emulsions is presented as a linear function, which suggests the constancy of structural parameters of the oil system and the absence of phase transitions in the above mentioned temperature range. The bond strength in associative complexes in each structural state corresponding to a certain temperature is characterized by the heat of activation of the viscous flow Eact. The values of Eact for 3 oil wells of the Usinskoye oilfield are 41.4 - 68.5 kJ/mol (Table 3).

Table 3. Dynamic viscosity and activation energy of oils and emulsions based on them

No.	Oil		10% emulsion		50% emulsion	
	η, Pa*s	Eact, kJ/mol	η, Pa*s	Eact, kJ/mol	η, Pa*s	Eact, kJ/mol
1	3.8	41.4	4.2	56.3	10.1	32.2
2	4.3	45.3	4.9	61.5	12.5	40.7
3	5.8	68.5	7.5	81.0	28.0	64.9

The maximum activation energy is characteristic of the oil sampled from the well No. 3. Introduction of water to the oil system up to 10% leads to an increase in Eact values by 16-27% compared to Eact of oil, which indicates to an increase in intermolecular interactions and an increase in the degree of the system associativity due to the formation of adsorption (solvation) sheaths consisting of resin-asphaltene components at the oil-water interface.

With a further increase in water cut to 50%, the Eact values decrease relative to the Eact of the 10%-th emulsion and oil. This is due to a significant decrease in the proportion of the oil phase in the composition of emulsion and in the content of emulsifying components, which is accompanied by a decrease in the strength of the formed associative combinations.

To assess the role of the dispersed phase in the formation of emulsions, the group composition of oils and intermediate layer of 10% emulsions was analyzed (Table 4). The formation of an intermediate layer during the separation of the emulsion is the result of incomplete destruction of the armor shells on droplets of emulsified water in oil during the retarded coalescence and fall of these drops.

Analysis of the group composition revealed that the Usinsk oil can be classified as a low-paraffin containing 1.8-2.1 wt% of solid paraffin hydrocarbons and highly resinous oil. The content of asphaltenes (A) in the oil of the Usinskoye oilfield sampled from various wells is approximately the same and amounts to 7.0 - 7.7 wt %. The total mass fraction of benzene (BR) and alcohol-benzene resins (ABR) varies significantly: 12.6 wt % in the well No. 1 and 22.1 wt % in the well No. 3. The largest amount of resin-asphaltene components was found in the oil from the well No. 3, which explains its high density and viscosity. The proportions of saturated (SHs) and aromatic hydrocarbons (AHs) are the highest in the well No. 1.

Table 4. Composition of the oil from the Usinskoye oilfield and that of intermediate layers of oil-water emulsions

Sample	Content, wt%					
	n-alkanes	SHs	AHs	BR	ABR	A
Well 1	1.9	61.5	17.0	5.4	7.2	7.0
i/l of the well 1	1.2	51.1	15.0	10.3	13.5	8.9
well 2	1.8	56.9	18.0	6.3	9.3	7.7
i/l of the well 2	1.5	45.0	17.4	9.3	15.3	9.3
well 3	2.1	52.6	15.6	9.5	12.6	7.6
i/l of the well 3	1.9	47.4	13.3	11.2	16.4	9.8

It is evident from Table 4 that an increase in the content of resin-asphaltene components is observed in the group composition of the intermediate layers isolated from 10% oil-water emulsions. Hence, the content of asphaltenes has increased by 18 - 23% and that of alcohol-benzene and benzene resins by 20 - 50%. The maximum increase in resin-asphaltene components reaching 70% was observed in the intermediate layer of emulsion of oil from the well No. 1, which was due to the increase in the proportion of resins.

4. Conclusion

By the example of similar in oil composition and physicochemical properties high-viscosity oil sampled from three wells at the Usinskoye oilfield, it has been shown that the stability of the emulsion and the strength of supramolecular structures depend on the component composition of the oil. The appearance of water in the oil from the well 1 containing the smallest amount of resin-asphaltene components compared to other wells leads to a significant increase in the proportion of resins in the composition of the intermediate layer. Moreover, 10% emulsions formed from this oil are characterized by minimum viscosity values and a maximum average droplet diameter. The increase in the proportion of resins and asphaltenes in the oil well 3 leads to the formation of emulsions characterized by higher values of the activation energy of viscous flow and viscosity, which is indicative of the formation of more stable supramolecular systems in the objects under study. At the same time, a high content of resin-asphaltene components promotes the formation of emulsions with a minimum droplet diameter in comparison with other emulsions based on oil sampled from the wells 1 and 2.

References

[1] Wong S F, Lim J S and Dol S S 2015 *J. Petrol. Sci. Eng.* **135** 498
[2] Mousavi M, Abdollahi T, Pahlavan F and Fini El. H. 2016 *Fuel* **183** 262
[3] Shi Ch, Zhang L, Xie L, Lu X, Liu Q, He J, Mantilla C A, Van den berg F G and Zeng H 2017 *Langmuir* **33** 1265
[4] Speight J G J. 1999 *J. Petrol. Sci. Eng.* **22** (1-3) 3
[5] Spiecker M, Gawrys K L, Trail C B and Kilpatrick P K 2003 *Colloids and Surf.,* A **220** 9
[6] Nebogina N A, Prozorova I V, Yudina N V and Savinykh Yu V 2010 *Petroleum Chemistry* **50** (2) 158
[7] Sullivan A P, Zaki N N, Sjöblom J and Kilpatrick P K 2007 *Can. J. Chem. Eng.* **85** (6) 793
[8] Peralta-Martinez M V, Arriola-Madellin A, Manzanares-Papayanopoulos E, Sanchez-Sanchez R and Palacios-Lozano E M 2004 *Petroleum Science and Technology* **22** (7-8) 1035
[9] Evdokimov I N, Eliseev N Yu and Eliseev D Yu 2004 *Fuel* **83** (7-8) 897
[10] Loskutova Y V, Yudina N V, Volkova G I and Anufriev R V 2017 *International Journal of Applied and Basic Research* **10-2** 221.
[11] Volkova G I and Yudina N *V* 2017 *South Siberian Scientific Bulletin* **4** (20) 282

IOP Conf. Series: Materials Science and Engineering **696** (2019) 012015 doi:10.1088/1757-899X/696/1/012015

Algorithms and models of multirobot systems and their implementation in ROS

M V Shikhman and S V Shidlovskiy

National Research Tomsk State University, Tomsk, Russia

E-mail: Shikhmar@gmail.com

Abstract. Conducting research in the Arctic takes place in extreme conditions. In these conditions, it's effective to use multirobot systems that can function for days under any weather conditions. The article discusses the basic principles of multirobot control systems. Mathematical models for the formation of control actions of multirobot are obtained. The basic methods for modeling multi-robot systems in ROS are described.

1. Introduction

The development of robotics allows the use of robotic systems in all spheres of human activity. Robotic systems are a combination of sensory, actuating and microcontroller devices. Their collaboration allows us to achieve various goals and to solve a wide range of technological problems. The use of robots is especially effective in places inaccessible to humans, as well as when performing work associated with risks to human life and health. Robotic systems can work around the clock in all weather conditions. These advantages are especially important when carrying out work in the Arctic, and that is why achievements in this area play an important role in the effective development of the Arctic lands.

2. Multirobot control systems

Even greater efficiency can be achieved with groups of robots. Multirobot systems (MRS) allow you to expand the radius of the system, to achieve your goals better because of the distribution of tasks. However, it is obvious that all this entails an increase in complexity in the development and operation of such systems.

Multirobot system is a group of mobile robots that coordinate their actions to achieve a common goal [1]. Thus, the task of multirobot control system is to form control actions for each robot, ensuring optimal achievement of the goal. The system control scheme looks as follows (Figure 1). The system has a specific goal, the multirobot control system generates control signals for each of the robots in accordance with some algorithms. In addition, this system must take into account the state of the external environment for the formation of control actions. The received signals in some way change the state of each of the robots, and as a result, robots also interact with the environment too. Information about changes in the state of the group, as well as changes in the environment should return to the control system. This is necessary in order to understand whether this goal has been achieved and what needs to be done next to achieve it.

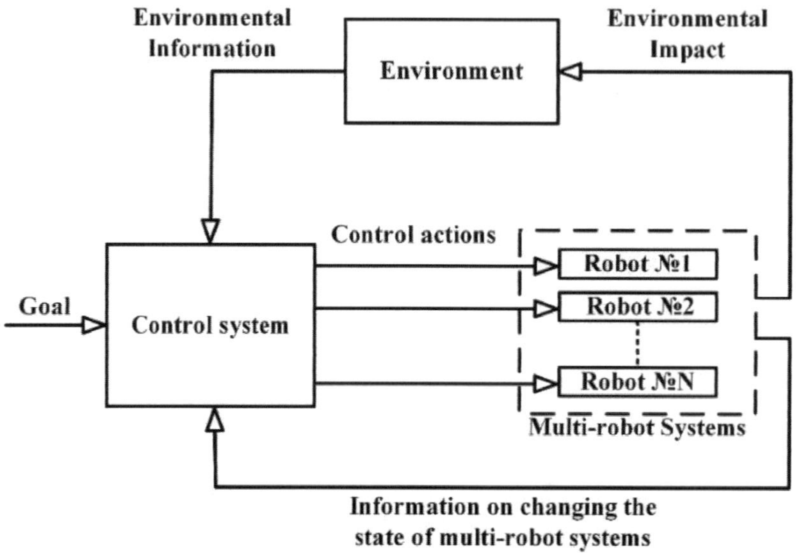

Figure 1. Multirobot control systems.

3. Coordination strategies for multirobot systems

The operation of the multirobot control system can be based on various control algorithms. Depending on this, the following strategies are can be used: centralized MRS and decentralized MRS, which is divided into collective and swarm MRS (Figure 2) [2].

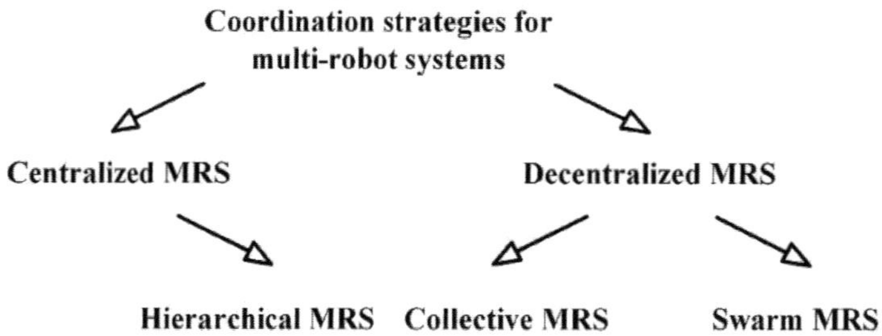

Figure 2. Coordination strategies for multirobot systems.

The strategy of centralized coordination implies the presence of some central control device, which completely takes all the necessary decisions to achieve the goal, and also forms the control actions for all robots (Figure 3).

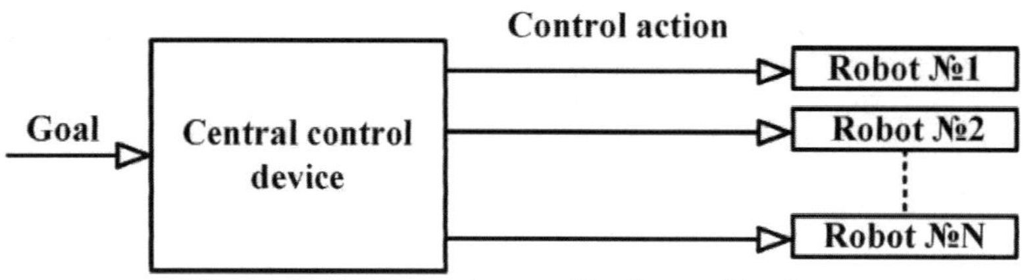

Figure 3. Strategy of centralized coordination.

A type of centralized MRS is hierarchical MRS. The peculiarity of hierarchical coordination is that the central device transfers control actions to a limited group of control devices (central devices of the second level), and it already generates control actions for all the remaining robots. The central devices of the second level can serve as other robots. In addition, there may be several levels of hierarchical access (Figure 4).

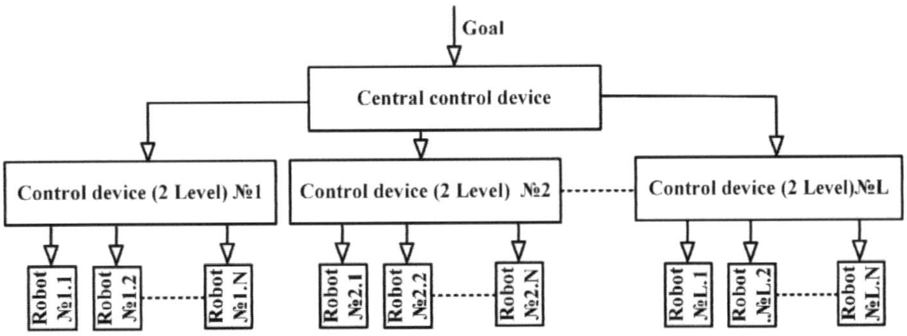

Figure 4. Strategy of hierarchical coordination.

The essence of a decentralized coordination strategy is that each robot in a group independently makes decisions about its actions. In the case of swarm MRS, robots make decisions only based on the purpose and interactions with the environment (Figure 5). If the coordination is collective, the robots interact with each other to make more effective decisions (Figure 6). Thus, the difference between swarm and collective coordination is the presence of some kind of information exchange channel between robots.

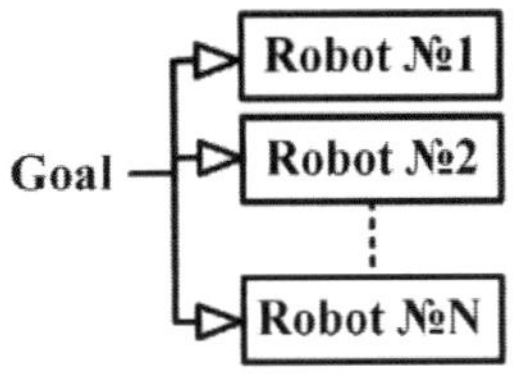

Figure 5. Strategy of swarm coordination.

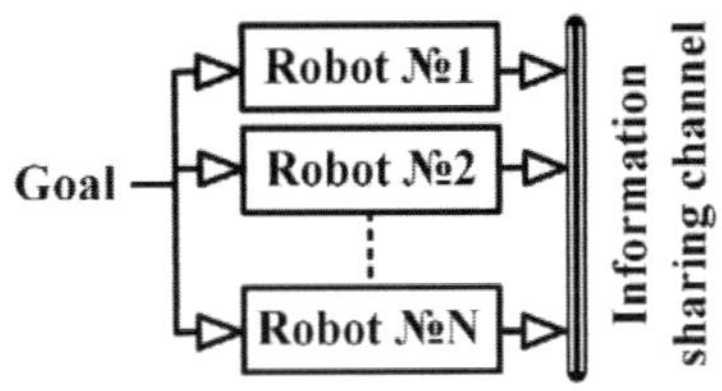

Figure 6. Strategy of hierarchical coordination.

4. Mathematical model for the formation of control actions for robots

Consider the formation of control actions for moving in multirobot system. With a central coordination strategy, the central control device or central robot can set the final goal in front of the robot, that is, set the coordinates of movement. In this case, the formation of control actions is not difficult and can be considered as the formation of control actions for a single robot. Another situation where the movement of robots in a group depends on the movements of the central robot. In this case, consider the central robot (the leader robot) and the robots following it as material points on the plane.

Let the central robot at some point in time have the coordinates (x_c, y_c). The remaining robots of the group, which consists of i robots, will have coordinates (x_i, y_i) and an angle of rotation relative to a rectilinear trajectory θ_i. The distance between the central and i-th robots will be equal to x_{di} and y_{di} along the corresponding axes (Figure 7).

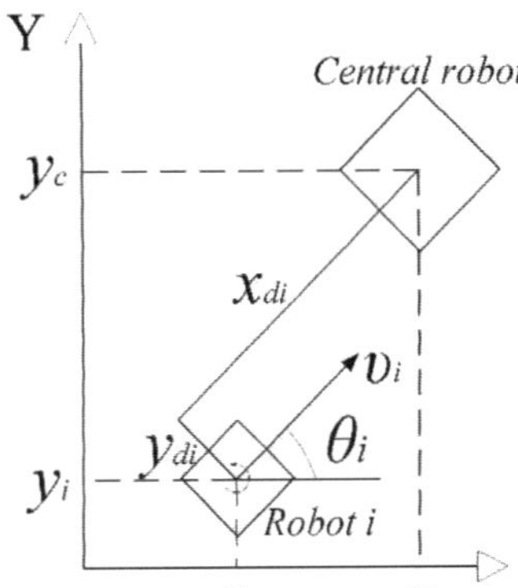

Figure 7. Robot kinematics model.

Then the system that combines all these parameters will look like [3]:

$$\begin{bmatrix} x_c \\ y_c \\ \theta_c \end{bmatrix} = \begin{bmatrix} x_i + x_{di}\cos\theta_i - y_{di}\sin\theta_i \\ y_i + x_{di}\sin\theta_i + y_{di}\cos\theta_i \\ \theta_i \end{bmatrix}.$$

We define a kinematic model of the motion of the central robot. Let us denote the linear speed of the robot υ_c and its rotation speed ω_c. At the initial time, the robot has coordinates y_c, x_c. Then, when the robot rotates by a certain angle θ_c, the following differential expressions can be written:

$$\begin{bmatrix} \dot{x}_c \\ \dot{y}_c \\ \dot{\theta}_c \end{bmatrix} = \begin{bmatrix} \upsilon_c \cdot \cos\theta_c \\ \upsilon_c \cdot \sin\theta_c \\ \omega_c \end{bmatrix},$$

where θ_c is a set of pairs of closest points; $\dot{x}_c$ – x-axis instantaneous speed; $\dot{y}_c$ – y-axis instantaneous speed.

Combining the two data systems, we get:

$$\begin{bmatrix} \dot{x}_c \\ \dot{y}_c \\ \dot{\theta}_c \end{bmatrix} = \begin{bmatrix} B_i \\ B_\theta \end{bmatrix} \begin{bmatrix} \upsilon_i \\ \omega_i \end{bmatrix},$$

where

$$B_i = \begin{bmatrix} \cos\theta_i & -x_{di}\sin\theta - y_{di}\cos\theta_i \\ \sin\theta_i & x_{di}\cos\theta_i - y_{di}\sin\theta_i \end{bmatrix},$$

$$B_\theta = [0 \quad 1].$$

Knowing the coordinates of each of the robots, you can calculate the length of the vector between them. For the i-th robot [4]:

$$L_i = \sqrt{(x_c - x_i)^2 + (y_c - y_i)^2}.$$

When we need to move the robot, we ask him the coordinates in which he should be as a result of this movement. By changing its position in space, the central robot will change its coordinates, which will become $(x_{c1},\ y_{c2})$. Knowing the corresponding coordinates, we can find the motion vector by the formula:

$$\overline{R_c} = (x_c - x_{c1}, y_c - y_{c1}),$$

where $(x_c,\ y_c)$ – coordinates of the starting point in the time interval; $(x_{c1},\ y_{c1})$ – coordinates of the end point in the time interval.

Based on this, we can find the angle of this vector by the formula:

$$\theta_c = arctg\left(\frac{y_c - y_{c1}}{x_c - x_{c1}}\right).$$

The obtained result allows one to form a control action on the robotic engines for the group, that is, to set robots the angles of movement θ_i.

This approach can also be used in a strategy of collective coordination to avoid collisions between robots and increase the radius of the system by controlling the distance between robots.

5. The basics of implementing multirobot systems in ROS

Modeling is effectively used in the development of robotic systems; this allows us to evaluate the effectiveness of certain algorithms in various conditions. One of the most popular modeling tools is the ROS (Robot Operating System). Let us consider the basic principles of building multirobot systems in ROS.

5.1. Adding a multi-robot

IOP Conf. Series: Materials Science and Engineering **696** (2019) 012015 doi:10.1088/1757-899X/696/1/012015

The first thing you need when modeling a system in ROS is to place several robots in a production environment. The environment and robot simulator work through the World file, which contains a description of the scene for the simulation. This is because of the fact that you can add the desired number of robots to the scene. Let us consider this with the example of the ROS Stage Simulator. Stage simulates a population of mobile robots, sensors and objects in a 2D bitmapped environment. Adding the following lines to the world file, we get a simulation of three robots in one environment (Figure 8):

# robots
erratic(pose [-11.5 23.5 0 0] name "robot0" color "yellow")
erratic(pose [-13.5 23.5 0 0] name "robot1" color "red")
erratic(pose [-12.5 21.5 0 0] name "robot2" color "green").

Figure 8. Adding a multi-robot.

5.2. Robot control

The next task is to send control actions to robots. Suppose we need to signal a certain robot so that it goes straight. There are two ways to do this.

1. For example, to make robot 2 move forward type: rostopic pub /robot_2/cmd_vel -r 10 geometry_msgs/Twist '{linear: {x: 0.2}}'.

2. You can create a ROS node that will make one of the robots move forward. To do this, create a launch file and add the following code there:

<launch>
<node name="stage" pkg="stage_ros" type="stageros" args="$(find stage_multi)/world/willow-multi-erratic.world"/>
<node name="move_robot_2" pkg="stage_multi" type="move_robot" args="2" output="screen"/>
</launch>

The result will be the same in both cases (Figure 9).

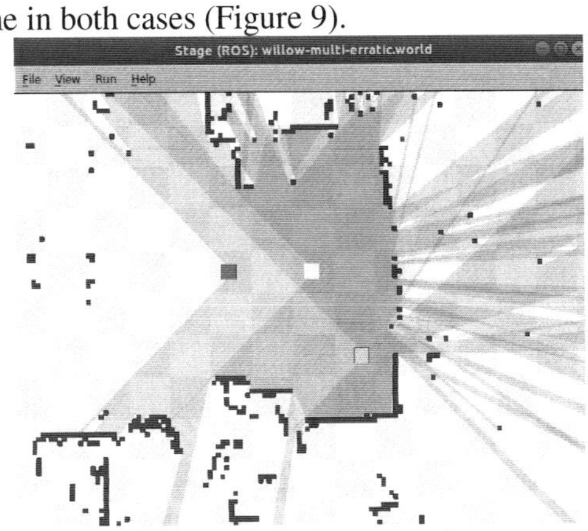

Figure 9. Robot control.

5.3. Coordinate between robots

IOP Conf. Series: Materials Science and Engineering **696** (2019) 012015 doi:10.1088/1757-899X/696/1/012015

When using the strategy of collective coordination, you need to coordinate and synchronize behavior between robots. For example, we create a condition under which robots begin to move only after they receive messages about the readiness of each of the robots. For this, it is necessary to create a shared topic in which robots will publish a ready status message; when such a message arrives from each robot, the monitoring node will publish a broadcast message stating that all robots are ready. Messages are described using msg files; the structure of readiness messages has the following form:

Header header
int32 robot_id
bool is_ready.

Next, we need to create the monitor node that will listen for the ready messages and announce when all robots are ready. As a result, we will be able to receive information about the readiness of robots (Figure 10). When all the robots are ready, they will begin to move (Figure 11).

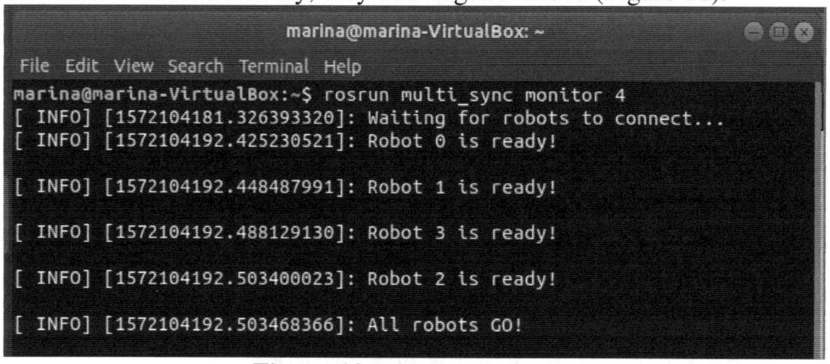

Figure 10. Robots are ready.

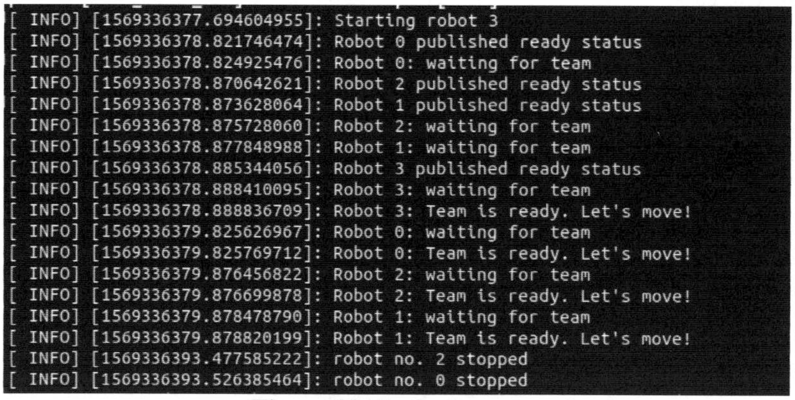

Figure 11. Begin to move.

6. Conclusion

Multirobot control systems are effective in extreme conditions of the Arctic. Different group control strategies allow you to use them for various purposes. One of the convenient software environments for implementing group control of robots is ROS, which allows both strategies of centralized and decentralized coordination.

Acknowledgments

The research was carried out in Tomsk State University with the financial support of the Ministry of Science and Higher Education of the Russian Federation, a project «Development of an intelligent control system for a group of robotic machines in the tasks of the agro-industrial complex»

The authors are grateful to Tatiana B. Rumyantseva from Tomsk State University for English language editing.

References

[1] Zhi Yan, Nicolas Jouandeau, Arab Ali Cherif 2013 *International Journal of Advanced Robotic Systems* **10(12)** doi 10.5772/57313
[2] Kalyayev I A 2001 *Modeli i algoritmy kollektivnogo upravleniya v gruppakh robotov* (Moscow: Fizmatlit) p 280
[3] Toru Namerikawa, Chika Yoshioka 2008 *IFAC Proceedings Volumes* pp 5149-5154 doi 10.3182/20080706-5-KR-1001.00865
[4] Ilyichev K V, Mantserov S A 2017 *Vestnik PNIPU* **21** pp 91-108

Wall ceramic products based on aluminum silicate materials

N K Skripnikova[2] *, A I Potekaev[1], V V Shekhovtsov[2], G G Volokitin[2], M A Semenovykh[2] and O G Volokitin [2]

[1]Tomsk State University, Tomsk, Russia
[2]Tomsk State University of Architecture and Building, Tomsk, Russia

*E-mail: nks2003@mail.ru

Abstract. The paper focuses on wall ceramic products modified by vitrified aluminosilicate microspheres. It is shown that during phase transformations occurred in firing polymineral clays and vitrified microspheres, the formation of mullite- and anorthite-like compounds and gehlenite is observed within the 950–1050°C temperature range. The formation of such phases provides a synthesis of a strong aluminosilicate scaffold for ceramic materials.

1. Introduction
New standards for thermal protection of buildings require the creation of new wall materials possessing improved thermophysical properties. It is therefore necessary to develop construction materials, which would meet the specific requirements for their application in severe climatic conditions.

The development and implementation of advanced ceramic wall materials having high mechanical-and-physical and thermophysical properties with the addition of industrial waste, is an urgent challenge for the modern materials science in construction. The presence in ashes of vitrified microspheres with a spectrum of the most important properties, allows them to be used in the production of wall materials.

The integrated use of clay materials in compositions with vitrified microspheres allows producing fired wall materials with high mechanical-and-physical and thermophysical properties and low thermal conductivity [1–5].

The aim of this work is the development of compositions for ceramic construction materials based on different clay minerals and vitrified microspheres using justified conditions of the structure formation.

2. Experimental
Natural and man-made raw materials are studied to synthesize ceramic wall products. Clays from Kemerovo, Novosibirsk and Tomsk regions are used as natural materials for the manufacture of fired wall materials.

A large amount of waste has accumulated in the Siberian region. Most waste is produced by mining, metallurgical and heat power industries [6]. On the territory of the Kemerovo region, several ash-disposal areas store hundreds of thousands of tons of waste. Vitrified ash microspheres are chosen as a research object.

Sampling materials from four mineral deposits were explored, namely: Rodionovskoe in the Tomsk region, Belovskoe and Egozovskoe in the Kemerovo region and Verkhnii Koen in the Novosibirsk region. Their chemical, mineral and grain-size compositions were studied to evaluate the possibility of application in manufacturing ceramic construction materials.

Table 1 summarizes the chemical composition of clay materials. The oxide content includes 12.75–14.17 % aluminum oxide, 3.33–5.56 % iron oxide, 4.3–9.10 % calcium oxide, and 0.50–2.9 % magnesium oxide.

Table 1. Chemical composition of clay materials.

Deposits	Oxide content, wt.%							
	SiO_2	Al_2O_3	Fe_2O_3	CaO	MgO	R_2O	TiO_2	Ignition loss
Belovskoe	58.80	13.77	5.56	9.10	0.50	1.09	-	11.18
Egozovskoe	66.44	12.75	4.36	6.25	1.4	1.88	-	6.92
Verkhnii Koen	65.2	12.9	4.9	4.3	2.9	4.1	0.8	4.9
Rodionovskoe	66.0	14.17	3.33	5.2	2.08	2.67	0.63	7.26

Using the data from Table 1, the clay materials can be grouped according to their industrial applications as depicted in Fig. 1. The mole ratio between Al_2O_3 and SiO_2 and a sum of moles of fusing agents are accepted as clay material indicators.

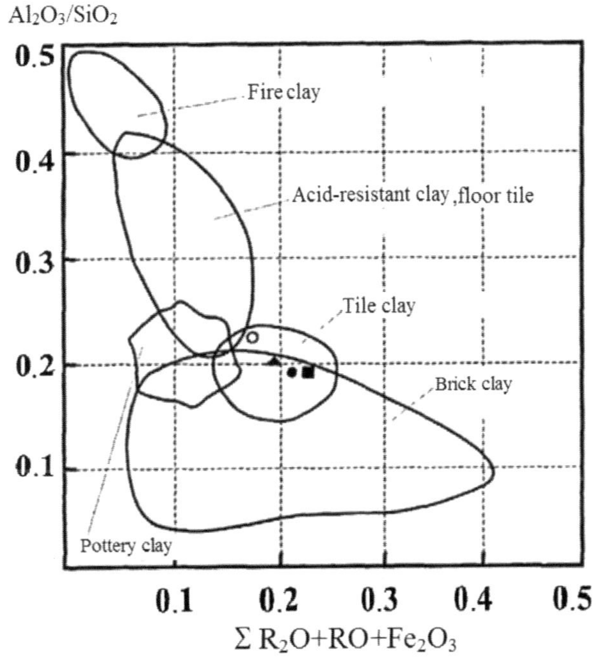

Figure 1. Industrial application of clays depending on their chemical composition: • – Egozovskoe deposit; ○ – Belovskoe deposit; ■ – Verkhnii Koen deposit; ▲ – Rodionovskoe deposit.

As can be seen from Fig. 1, clays from Belovskoe deposit can be used in the production of clay tiles, whereas clays from Egozovskoe, Rodionovskoe and Verkhnii Koen deposits can be used in both brick and clay tile production.

In terms of the grain-size composition, all the clays relate to a coarsely dispersed raw material. Moreover, they have a low content (less than 1%) of coarse-grained inclusions. These clays can be used in the brick and clay tile production. The low degree of dispersion provides these clays with low plasticity and sensitivity to dehydration. The plasticity index of clays varies from 8.0 of Verkhnii Koen deposit to 12.25 of Belovskoe deposit. The dehydration sensitivity coefficient ranges between 1.026 and 1.31. Clays from all the four deposits possess a medium plasticity, and the dehydration sensitivity coefficient is nearly 1.

In Fig. 1 imaging points of the clay compositions lie in the elementary triangle of anorthite–cristobalite–silica, with the eutectic point at 1100 °C.

Kaolinite–montmorillonite composition is favorable for sintering clay minerals as it is characterized by a wider increase interval of the melt.

Vitrified microspheres are considered as a man-made raw material which remains after coal burning in furnaces and is stored in ash-disposal areas using a hydraulic method.

Table 2 presents the chemical composition of vitrified microspheres, which shows that it is an aluminosilicate material. A significant amount of aluminum oxide, fusing agents and iron oxide promotes the formation of both liquid and crystal (mullite, fayalite) phases, which improve the product strength. The presence of carbon residue and fusing agents lowers the temperature of the liquid phase formation and saves the fuel.

Table 2. Chemical composition of vitrified microspheres

Oxide content, wt.%						
SiO_2	Al_2O_3	FeO	CaO	MgO	R_2O	Total
62.80	24.40	5.04	4.60	1.82	1.34	100

The mineralogical composition of vitrified microspheres shows that the amount of amorphous aluminosilicate and crystalline materials is 91 % and 9 %, respectively. The crystal phase consists of 84–85 % of silica and 15–16 % of mullite and hematite.

The obtained mineralogical composition is supported by the X-ray diffraction (XRD) patterns presented in Fig. 2.

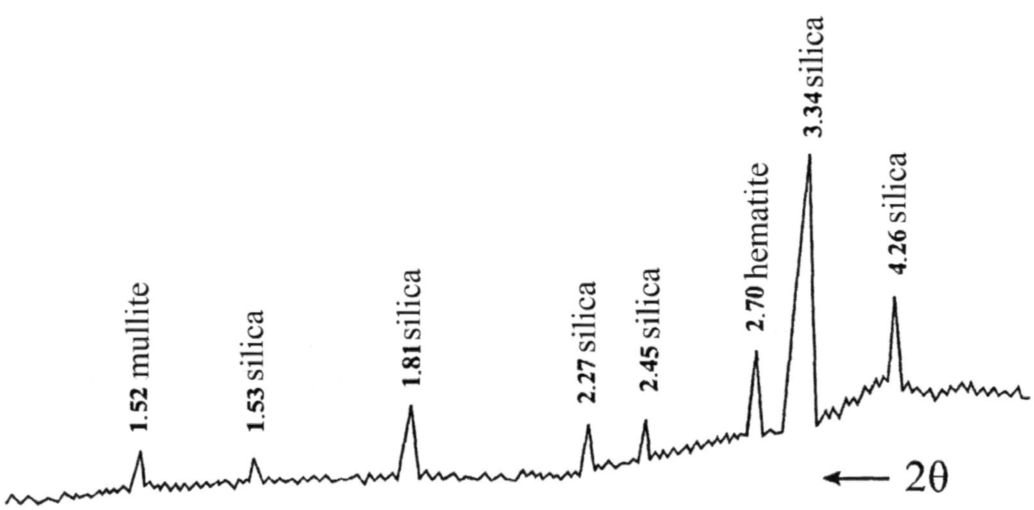

Figure 2. XRD patterns of vitrified aluminosilicate microsphere.

According to our morphological observations, the shape of vitrified microspheres is fully spherical due to their formation from the gas-containing melt [7].

It should be noted that the chemical and mineralogical compositions of vitrified microspheres depend on the ash composition and is determined by the composition of the initial fuel (coal, shale) they are produced from [8].

A study of the properties of vitrified microspheres shows that they mostly consist of hollow spherical particles. Therefore, the properties of the end composite product will depend on the shape and properties of vitrified microspheres.

The great difference in the bulk density of vitrified microspheres and clay material can be explained not only by their dispersion degree and mineralogical composition, but above all, by the fact that vitrified microspheres mainly consist of the vitreous phase. These microspheres are filled with flue gases comprising carbon oxide, an uncomplete combustion product.

In firing the ceramic material, gas in microspheres burns, thereby promoting the temperature increase inside the final ceramic material. Additionally, they consist of carbon particulates in the form of coke, the amount of which can be rather large.

Based on this information, it is hypothesized that the structure of the ceramic material generates from a mixture of clays and vitrified microspheres. At the initial stage of sintering, during the uncomplete combustion of residual fuel with CO formation, the created reducing conditions provide the transfer of iron oxide to ferrous iron in clay minerals and the temperature decrease of the primary melt due to the formation of low-temperature eutectics. This melt creates conditions for the intense sintering of interacting particles and the increase in the liquid phase amount due to the dissolution of the vitreous phase, feldspar and clays in the primary alkaline ferrous melt. The sintering process is accompanied by the pore formation conditioned by the liquid phase swelling by gaseous products resulting from burning of carbon oxide and carbon particulates and the removal of chemically-bound water from clay fractions. The appeared pores are mostly closed, with walls strengthened by wetting of the vitreous phase. The sintering process in the presence of the liquid phase intensifies the mullite crystallization, thereby hardening the obtained product due to a complete structure formation [9].

A composition of 80 wt.% clays and 20 wt.% aluminosilicate microspheres was used to investigate phase transformations in firing the ceramic material. The specimens 120×60×15 mm in size were prepared and fired at a temperature of 1000 and 1050 °C.

The XRD patterns of the fired clay and aluminosilicate microsphere composition is given in Fig. 3. One can see the formation of mullite crystals $3Al_2O_3 \cdot 2SiO_2$, which has a positive effect on the mechanical strength of the obtained ceramic composite.

The XRD analysis of the fired composition proves the presence of such crystal phases as silica, wollastonite, gehlenite, albite, cyanite, and mullite- and anorthite-like compounds.

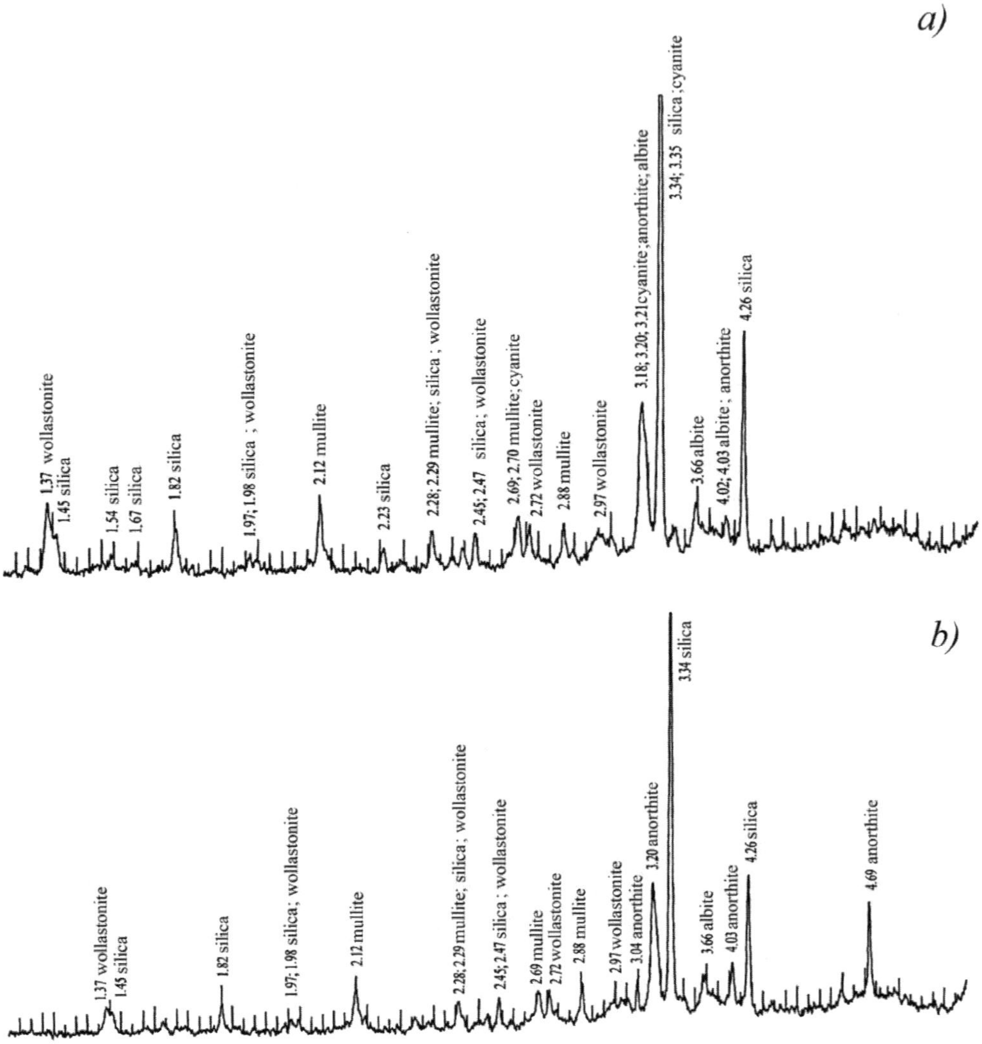

Figure 3. XRD patterns of specimens fired at 1000 °C: *a* – clays; *b* – clays and vitrified aluminosilicate microspheres in the ratio of 80:20.

Observations of the composite specimen microstructure performed on a scanning electron microscope (SEM) show that it is homogeneous, with both closed and interconnected pores (Fig. 4). There is a slight spread in values of the pore size. The shape and size of the structural elements in the obtained material are more regular than in conventional material with the granulated structure. In porous materials, the strength parameter is determined by the grain size, shape and size of interparticle necks, and their spatial orientation.

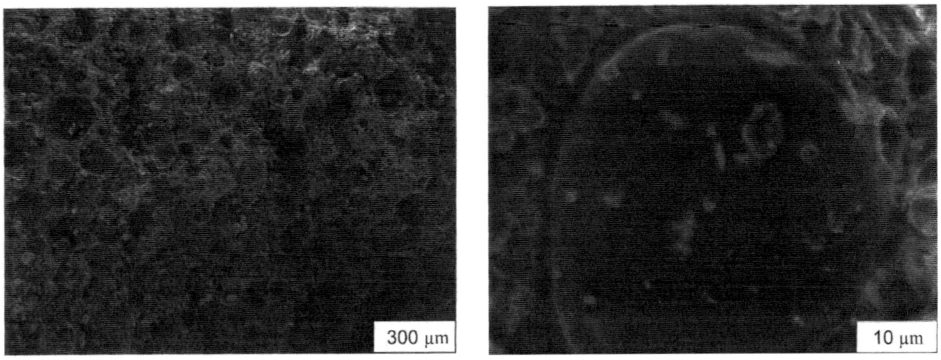

×100 ×2000

Figure 4. SEM images of the specimen consisting of clays and vitrified aluminosilicate microspheres.

In Fig. 4, the contact layer is well defined around the microspheres. It is formed during sintering the clay material and vitrified microspheres. This demonstrates that the obtained melt of the clay minerals envelopes the microsphere surface and partially hardens the aluminosilicate scaffold of the ceramic material.

3. Conclusions

Based on the investigation results of phase transformations occurred during firing polymineral clays and vitrified microspheres, it can be stated that mullite- and anorthite-like compounds and gehlenite formed within the temperature range of 950–1050°C. The formation of such phases provided a synthesis of a strong aluminosilicate scaffold for the ceramic material. The firing temperature of 1000 °C was detected to be enough for the production of the effective wall materials, because it provided the formation of mullite- and anorthite-like phases between the vitrified microspheres and clay minerals.

4. Acknowledgments

The reported research was funded by Russian Foundation for Basic Research and the government of the region of the Russian Federation, grant №18-48-703004.

References

[1] Ignat'ev G V. and Shushkov D A 2018 *Mineraly: stroenie, svoistva, metody issledovaniya* **9** 77–78 (in Russain).
[2] Zakarlyuka S G and Yatsenko N D 2016 *Inzhenernyi vestnik Dona* **3** 42 79 (in Russian).
[3] Ge X, Zhou M, Wang H, et al. 2019 Ceramics International **45** 9 12528–34 DOI: 10.1016/j.ceramint. 2019.03.190.
[4] Shishkin Аб Laksa A, Shodlovska V, et al. 2017 *Key Engineering Materials* **721** 316–321 DOI: 10.4028/www.scientific.net/KEM.721.316.
[5] Su Zh, Xi X, Hu Y, et al. 2014 *Journal of Porous Materials* **21** 5 601–609 DOI: 10.1007/s10934-014-9806-7.
[6] Abdrakhimov V Z, Lobachev D A, Abdrakhimova E S, et al. 2016 *Biosfernaya sovmestimost': chelovek, region, tekhnologii* **4** 16 31–38 (in Russian).
[7] Shekhovtsov V V, Volokitin O G, Volokitin G G, et al. 2018 *Glass and Ceramics* **75** 1–2 32–35.
[8] Vlasov V A, Shekhovtsov V V, Volokitin O G, et al. 2018 *Russian Physics Journal* **61** 4 708–714.

[9] Skripnikova N K, Litvinova V A, Volokitin G G, et al. 2017 Vestnik *TSUAB* **2** 141–147 (in Russian).

Methods for describing structure of novel porous materials: a review

A Yu Smolin[1,2], **G M Eremina**[1,2] **and S Yu Korostelev**[1]

[1] Institute of Strength Physics and Materials Science SB RAS, Tomsk, 634055, Russia
[2] National Research Tomsk State University, Tomsk, 634050, Russia

E-mail: asmolin@ispms.ru

Abstract. The physical and mechanical properties of heterogeneous materials are mainly determined by their structure. Due to the wide capabilities of modern technologies for the formation of complex multilevel pore structure of ceramics and other materials, the ability of modern methods to describe and characterize such a structure completely is of particular interest. Traditionally, the pore structure is characterized by a large number of parameters such as porosity, type and shape of pores, the fraction of cross-sectional area of channels, pore size distribution, etc. However, recently, at least three new quantities are often used to describe and measure the structure of various complex media: the Minkowski functionals, fabric tensors, and correlation functions. This review is aimed to consider all these three approaches and their capabilities for describing and measuring the peculiarities of the structure of porous materials.

1. Introduction

Due to the significant influence of the structure of the pore space on the strength and elastic properties of porous materials [1–7], as well as the wide capabilities of modern technologies for the formation of complex multilevel pore structure [8, 9], the ability of modern methods to describe and characterize such a structure completely is of particular interest. Traditionally, the structure of the pore space is characterized by the following parameters: porosity (volume fraction of all pores); its distribution in space; type of pores (open, closed, etc.); opening (the fraction of cross-sectional area of channels); shape and coefficient of pores tortuosity; pore size distribution; specific surface of pores; permeability and its distribution by filtration area [10]. Analysis of modern literature shows that at last decades three new approaches are also often used to describe and measure the structure of various complex media. The first approach is based on integral geometry and uses the Minkowski functionals and shapefinders. The second one uses so-called fabric tensors, which can be computed using both the geometric methods and the solid mechanics methods. The third approach is the most general and uses correlation functions. Below we consider all these three approaches and their capabilities for description and measuring the peculiarities of pore space structure of porous materials.

2. Minkowski functionals

In topology, the morphology of a three-dimensional structure can be completely determined by the following four quantities named Minkowski functionals (thereafter MFs) according to Hadwiger's theorem [11]

$$V_0 = \iiint dV\,;\, V_1 = \frac{1}{6}\iint dS;\, V_2 = \frac{1}{6\pi}\iint (\kappa_1 + \kappa_2)dS;\, V_3 = \frac{1}{4\pi}\iint \kappa_1\kappa_2 dS. \tag{1}$$

where κ_1 and κ_2 are the principal curvatures of the structure surface. Note, that the coefficients before integrals in (1) depends on the purpose of further use of the functionals values and could be normalized, for example, for getting Minkowski polynomials [12, 13]. The meanings of the first two functionals are very simple: V_0 is the volume; V_1 is proportional to the surface area of the structure. V_2 provides information about the shape and is proportional to the integral mean curvature; the meaning of V_3 is ambiguous (so-called Euler characteristic related to number of clusters minus number of tunnels plus number of cavities). MFs can be used to calculate such additional quantities, called "shapefinders", as the "thickness" T, the "breadth" B and the "length" L of the structure [14]

$$T = \frac{V_0}{2V_1};\ B = \frac{2V_1}{\pi V_2};\ L = \frac{3V_2}{4V_3}. \tag{2}$$

In addition, two dimensionless shapefinders "planarity" P and "filamentarity" F are used [14]

$$P = \frac{B-T}{B+T};\ F = \frac{L-B}{L+B}. \tag{3}$$

Many features of a 3D structure can be analysed using a plot of the filamentarity versus planarity. For example, the simplest structures are characterized by the following values: line— $P = 0$, $F = 1$; plane— $P = 1$, $F = 0$; sphere— $P = F = 0$.

Using a scanner or tomography, any structure can be represented as a grid of pixels or voxels. Results of the computer simulations are also obtained on grids, including uniform ones. Then one can use the Adler's algorithm to compute MFs in terms of such values as the number of grid nodes within the structure n_0; the number of edges n_1; the number of faces n_2; the number of cubic cells n_3; the total number of grid nodes N:

$$V_0 = n_3;\ V_1 = \frac{2(n_2 - 3n_3)}{9N};\ V_2 = \frac{2(n_1 - 2n_2 + 3n_3)}{9N^2};\ V_3 = \frac{n_0 - n_1 + n_2 - n_3}{N^3}. \tag{4}$$

In [12] the authors point out the importance of a morphological characterization of patterns in statistical physics. They illustrate the integral geometric approach to stochastic geometries by applying morphological measures to such diverse topics as percolation, complex fluids, and the large-scale galaxy distribution in the Universe. In particular, porous media may be generated by overlapping holes of arbitrary shape distributed uniformly in space. The percolation threshold of such porous media can be estimated accurately in terms of the morphology of the distributed pores. They also stress out that morphological measures are a novel method for the description of complex spatial structures aiming for relevant order parameters and structure information complement to correlation functions. Typical applications address Turing patterns in chemical reaction-diffusion systems, homogeneous phases evolving during spinodal decomposition, and the distribution of galaxies and clusters of galaxies in the Universe as a prominent example of a point process in nature.

In [13] the authors consider the effects of distortions (drift, noise, and blurring) on the morphological properties of complex random models, representative of a wide range of structure. Such distortions arise experimentally in imaging techniques due to diffraction, absorption and sample drift. The authors show how critically these distortions affect image quality as measured by the MFs.

In review [14] the authors give sufficient motivation for employing the Minkowski functionals in cosmology, they present a summary of the progress made in the direction of using the MFs to quantify the large scale structure by studying the geometry and topology of the superclusters and voids. Specifically, they stress the discriminatory power of MFs and of the derived morphological statistics, the shapefinders. Shapefinders are specifically important to study the shapes and sizes of the superclusters and voids. Several successes of shapefinders are highlighted there.

One of the important applications of using MFs to analyse complex structure belongs to computational fluid dynamics. For example, in [15] the authors consider an approach for studying the results of direct numerical simulation of large eddy simulations of multi-species turbulent mixing under high-pressure conditions.

Several metrics are used to understand the modelling results. First, a statistical analysis of the simulation results database for large-scale flow structures was performed to provide a metric for probing the accuracy of the proposed models as the flow fields obtained from accurate simulations should contain structures of morphology statistically similar to those observed in the filtered-and-coarsened simulations fields. To characterize the morphology of the large-scale structures, the Minkowski functionals of the iso-surfaces were evaluated for two different fields: the second-invariant of the rate of deformation tensor and the irreversible entropy production rate.

In [16] so-called anisotropic Minkowski functionals (AMFs) are used to capture local anisotropy while evaluating topological properties of the underlying grey-level structures. The authors evaluate the ability of this approach to characterize local structure properties of trabecular bone micro-architecture in ex vivo proximal femur specimens, as visualized on multidetector CT, for purposes of biomechanical bone strength prediction.

By virtue of their defining property of rotation-invariance, Minkowski functionals are not explicitly sensitive to directional and anisotropic features of morphology. This lack motivates the generalization to tensorial quantities, called Minkowski tensors. In [17] the authors provide a simple overview of the definition, properties and interpretation of Minkowski tensors, as well as advanced applications of these tensors to local anisotropy analysis of simple numerical models, to real-world data sets from scanning force microscopy of co-polymer films, X-ray tomography of granular matter and open-cell solid foams, and to defect detection in molecular dynamics simulations of metallic systems [18, 19].

It has to be noted, that Minkowski tensors can be also considered as a kind of fabric tensors, which are described in the next section. Good theoretical description of Minkowski functionals, their applications, as well as free software, are available online at https://morphometry.org/.

3. Fabric tensors

In modern scientific literature, the concept of fabric tensor is widely used to describe the structural features of various materials [20]. Apparently, it was initially used to describe the structure of geological media [21], then in medicine to study the structure of bone tissue [22–24], and, finally, in materials science for composites [17, 25]. Generally speaking, fabric tensors are understood as various tensor values that somehow characterize the structural sensitivity of a material. Fabric tensors aim at modelling through tensors both anisotropy and orientation of a material with respect to another one. These tensors can be seen as semi-global measurements since they are computed in relatively large neighbourhoods, which are assumed quasi-homogeneous. Therefore, in the review [20], the methods of the fabric tensor computation are classified into mechanics-based and morphology-based.

In its turn, mechanics-based methods can use the approach of solid mechanics; in this case, the simplest fabric tensor is the stiffness tensor. The second group of mechanics-based methods uses the wave propagation approach. Assuming poroelastic behaviour, it has been shown that wave propagation in trabecular bone, for example, can be characterized through the acoustic tensor, the solid-fluid interaction tensor, and the intrinsic permeability tensor, which describe the elastic and viscous effects in the media [24].

Morphology-based methods have two advantages compared to the mechanics-based methods. First, they are largely less computationally expensive than those obtained from mechanical simulations. Second, unlike methods based on mechanics, the resulting fabric tensors are not dependent on the boundary conditions applied during the simulations, homogenization schemes and/or general design of the simulations. However, as a counterpart, it is necessary to relate these fabric tensors with mechanical properties of the material, especially, elasticity. According to [20], a summary of the morphology-based methods is presented in Table 1.

Table 1. Summary of morphology-based methods to compute fabric tensors.

Boundary-based	Mean Intercept Length (MIL) tensor [26] Generalized MIL tensor [26] Global gradient structure tensor (GST) [22]
Volume-based	Volume orientation tensor (VO) [27] Star volume distribution (SVD) [27] Star length distribution (SLD) [27] Tensor scale [28] Inertia tensor [29] Sampling sphere orientation distribution (SSOD) [23]
Texture-based	Fractal dimension (FD) [30] Hurst orientation transform [31] Variance orientation transform [32] Line fraction deviation [33] Spatial autocorrelation [34] Different statistics [35]
Alternative methods	Minkowski tensors [17] Diffusion tensor imaging (DTI) [36] Texture tensor [37] Skeleton-driven [38] Assessment of the power spectrum [39]

It is important to remark that fabric tensors are not global measurements. Thus, it is possible to obtain fields of fabric tensors where tensors are computed locally with respect to a neighbourhood. However, large neighbourhoods are usually used, since regions of interest are usually assumed homogeneous. This imposes the problem of determining the appropriate size of neighbourhoods.

4. Correlation functions

In [40], it is noted that the total amount of information contained in the Minkowski functionals is much less than is necessary to describe even a relatively simple structure. At the same time, there is a method that allows one to quantitatively describe the internal structure of an object, and it is based on the calculation of the so-called correlation functions [41–44].

Each correlation function describes the probability of a certain image configuration, for example, that the points at the ends of an arbitrary segment lying on the image being studied are in the same phase. The most complete description of the correlation functions is presented in the book [42]. The simplest type of correlation function (*n*-point correlation function) shows the probability of finding *n* points in the same phase. The value of the one-point correlation function is equal to the volume fraction of the binary fraction (for example, porosity). The two-point correlation function is defined as the probability of simultaneously landing points $\boldsymbol{x}_1$ and $\boldsymbol{x}_2$ in the same phase (pores or solid phase of a porous material) and is the most studied function for describing random media.

Concerning heterogeneous materials, any effective properties, $\boldsymbol{K}_e$, is defined by a linear relationship between an average of a generalized local flux $\boldsymbol{F}$ and an average of a generalized local (or applied) intensity $\boldsymbol{G}$, i.e.,

$$\boldsymbol{F} \propto \boldsymbol{K}_e \cdot \boldsymbol{G} \tag{5}$$

As it is shown in [42], there are a variety of different correlation functions that naturally arise when the averaging process involved in relation (5) is rigorously carried out. Roughly speaking, the averaging process results in integrals in which the relevant local fields are weighted with the *n*-point correlation functions. More precisely, the averages are functionals of the *n*-point correlation functions.

In [40], it is also noted that for describing and restoring heterogeneous media there are not enough two-point correlation functions and under the prevailing conditions the most correct way of improving the accuracy of structure description could be based on using additional low-order functions ($n \leq 2$). Each of them represents the probability that the position of the endpoints of a segment or a segment itself must satisfy certain conditions. The following such functions have been proposed so far:

- cluster function (the segment ends are inside one cluster) [45];
- linear path function (the whole segment is in one phase) [46];
- chord length function [47];
- various functions of surfaces, for example surface-void and surface-surface functions [48];
- pore-size function [42].

At present, it is unknown which set of correlation functions is universal and sufficient for an accurate description of the structure and its properties for each specific case. Nevertheless, it is possible to increase the amount of information about the structure provided by the set of correlation functions by increasing their types and quantities. Correlation functions are a potential way to describe any structure because they allow correct solving the inverse problem and reconstructing the structure by the values of the complete set of correlation functions [40].

It is important that the correlation functions allow us to estimate how much information they can provide about a given structure. Such an estimate can be made by studying the parameters of the reconstruction algorithm and the number of confluent states when the same set of correlation functions corresponds to different structures [42, 43].

5. Conclusions

Review of the modern scientific literature shows that for a complete description of the complex structure the most powerful and mathematically correct are the use of Minkowski tensors and correlation functions. Fabric tensors are often actually the results that could be got after the structure analysis (at least for mechanics-based methods). The main advantage of the use of Minkowski functionals is that the corresponding free software is available with examples of its application. As a disadvantage of the use of correlation functions, we consider that their methods are based on random algorithms that seem to be very consuming in terms of computational time.

Acknowledgments

The study has been carried out at financial support of the Fundamental Research Program of the Russian Academy of Sciences for 2013–2020 (project III.23.2.3) and the Tomsk State University competitiveness improvement programme (grant 8.2.19.2018).

References

[1] Konovalenko Ig S, et al 2009 *Tech. Phys.* **54** 758–61
[2] Smolin A Yu, et al 2014 *Eng. Fract. Mech.* **130** 53–64.
[3] Kalatur E, Buyakova S, Kulkov S and Narikovich A 2014 *AIP Conf. Proc.* 1623 225–8
[4] Smolin I Yu, et al S 2016 *Proc. Struct. Integr.* **2** 3353–60
[5] Grigor'ev M V, et al 2017 *Tech. Phys. Lett.* 43 723–6
[6] Kulkov S N, et al 2015 Epitoanyag-JSBCM **67** 155–8
[7] Grigoriev M V, et al 2018 Epitoanyag-JSBCM 70 18–22
[8] Zhukov I, Buyakova S P and Kulkov S S 2016 Epitoanyag-JSBCM **68** 74–9
[9] Kulbakin D E, et al 2019 *IOP Conf. Ser.: Mater. Sci. Eng.* **511** 012006
[10] Rouquerol J, et al 1994 *Pure Appl. Chem.* **66** 1739–58
[11] Klain D A and Rota G-C 1997 *Introduction to geometric probability* (Cambridge University Press, Cambridge)
[12] Mecke K 1998 *Int. J. Modern Phys.* B **12** 861–99
[13] Arns C H, Knackstedt M A and K Mecke 2010 *J. of Microscopy* **240** 181–96
[14] Sheth V and Sahni V 2005 *Current Science* **88** 1101–16
[15] Borghesi G and Bellan J 2015 *Phys. Fluids* **27** 035117
[16] Nagarajan M B, De T, Lochmüller E-M, Eckstein F and Wismüller A 2014 Using anisotropic 3D

Minkowski functionals for trabecular bone characterization and biomechanical strength prediction in proximal femur specimens *Medical Imaging 2014: Biomedical Applications in Molecular, Structural, and Functional Imaging Proc. SPIE 9038* ed R C Molthen, J B Weaver (SPIE Medical Imaging, San Diego, CA) 903820

[17] Schröder-Turk G E, et al 2011 *Adv. Mater.* **23** 2535–53

[18] Golovnev I F, Golovneva E I and Fomin V M 2008 *Phys. Mesomech.* **11** 19–24

[19] Golovnev I F, et al 2015 *Phys. Mesomech.* **18** 179–186

[20] Moreno R, et al 2014 Techniques for computing fabric tensors: a review *Visualization and processing of tensors and higher order descriptors for multi-valued data, mathematics and visualization* ed C-F Westin, et al (Springer-Verlag, Berlin) 271–92

[21] Launeau P and Robin P Y F 1996 *Tectonophysics* **267** 91–119

[22] Tabor Z and Rokita E 2007 *Bone* **40** 966–72

[23] Varga P and Zysset P K 2009 *Med. Image. Anal.* **13** 530–41

[24] Cowin S C and Cardoso L 2011 *Biomech. Model Mechanobiol.* **10** 39–65

[25] Voyiadjis G Z and Kattan P I 2006 *Advances in Damage Mechanics: Metals and Metal Matrix Composites with an Introduction to Fabric Tensors* (Elsevier, Oxford)

[26] Moreno R, Borga M and Smedby Ö 2012 *Med. Phys.* **39** 4599–612

[27] Odgaard A 1997 *Bone* **20** 315–28

[28] Xu Z, Saha P K and Dasgupta S 2012 *Comput. Vis. Image. Underst.* **116** 1060–75

[29] Vasilić B, Rajapakse C S and Wehrli F W 2009 *Med. Phys.* **36** 3280–91

[30] Majumdar S, et al 1999 *Med. Phys.* **26** 1330–40

[31] Podsiadlo P, Dahl L, Englund M, Lohmander L S and Stachowiak G W 2008 *Osteoarthr. Cartil.* **16** 323–9

[32] Wolski M, Podsiadlo P, Stachowiak G, Lohmander L and Englund M 2010 *Osteoarthr. Cartil.* **18** 684–90

[33] Geraets W G M, et al 2008 *J. Biomech.* **41** 2206–10

[34] Wald M J, Vasilić B, Saha P K and Wehrli F W 2007 *Med. Phys.* **34** 1110–20

[35] Guggenbuhl P, Chappard D, Garreau M, Bansard J Y, Chales G and Rolland Y 2008 *Eur. J. Radiol.* **67** 514–20

[36] Capuani S, Rossi C, Alesiani M and Maraviglia B 2005 *Solid State Nucl. Magn. Reson.* **28** 266–72

[37] Graner F, Dollet B, Raufaste C and Marmottant P 2008 *Eur. Phys. J. E: Soft Matter Biol. Phys.* **25** 349–69

[38] Kinney J H, Stölken J S, Smith T, Ryaby J T and Lane N 2005 *Bone* **36** 193–201

[39] Brunet-Imbault B, Lemineur G, Chappard C, Harba R and Benhamou C L 2005 *BMC Med. Imaging* **5** 4

[40] Karsanina M V 2016 *Modeling and reconstruction of structure and properties of porous media using correlation functions* PhD thesis (Institute of Geosphere Dynamics of Russian Academy of Sciences, Moscow) [in Russian]

[41] Torquato S 1991 *Appl. Mech. Rev.* **44** 37–76

[42] Torquato S 2002 *Random Heterogeneous Materials: Microstructure and Macroscopic Properties* (Springer-Verlag, New York)

[43] Gommes C J, Jiao Y and Torquato S 2012 *Phys. Rev. E Stat. Nonlin. Soft Matter Phys.* **85** 051140

[44] Gommes C J, Jiao Y and Torquato S 2012 *Phys. Rev. Lett.* **108** 080601

[45] Torquato S, Beasley J D and Chiew Y C 1988 *J. Chem. Phys.* **88** 6540

[46] Lu B and Torquato S 1992 *Phys. Rev.* A. **45** 922

[47] Torquato S and Lu B 1993 *Phys. Rev.* E. **47** 2950

[48] Jiao Y, Stillinge F H r, Torquato S 2007 *Phys. Rev.* E. **76** 031110

IOP Conf. Series: Materials Science and Engineering **696** (2019) 012018 doi:10.1088/1757-899X/696/1/012018

The effect of mechanical treatment of TiNi powder on structure and phase composition

E V Abdulmenova and S N Kulkov

National Research Tomsk State University, Tomsk, Russia
Institute of Strength Physics and Materials Science SB RAS, Tomsk, Russia
National Research Tomsk Polytechnic University, Tomsk, Russia

E-mail: Ekaterina.V.Abdulmenova@yandex.ru

Abstract. In this work, the study of the effect of mechanical treatment of TiNi powder on structure and phase composition is presented. The powder was processed in high-energy planetary ball mill for different time. It has been shown that the average particle size measured by scanning electron microscopy changes not monotonous after mechanical treatment, but form of particles changes. The coherently diffracting domain size decreases from 43 nm to 5 nm with increasing of mechanical treatment time that significantly less than the average particle size measured by laser diffraction method and the average particle size calculated from the obtained values of the specific surface. That is to say, the powder particles consist of several crystallites. The width of all peaks changes after 10 s of mechanical treatment that can correspond to an increasing of the defect density. The lattices parameters of all phases change not more than 0.8 % after mechanical treatment. Two processes compete during mechanical treatment, characterizing either brittleness or plasticity. Many defects appear during mechanical treatment stipulate a quasi-amorphous state with increasing of mechanical treatment time.

1. Introduction

The creation of new materials is actual task for the development of new technologies used in Siberia and the Arctic region.

As it well known that amorphous phase, metastable crystalline and quasicrystalline phases, supersaturated solid solutions and nanostructuresa [1-3] can formed in result mechanical treatment and properties of materials can change [4, 5]. Mechanical treatment allows creating active states in a solid that can leads to accelerating of chemical reactions. It is crucial in use in different areas, for example, for the storage of hydrogen. Among hydride forming system which applied in this area, special emphasis should be placed on the intermetallic compound Ti- at near-equiatomic composition which as a rule has a complex phase content consisting with TiNi (B2), (B19'), Ti_2Ni ($E9_3$) and $TiNi_3$ ($D0_{24}$) [6, 7]. Therefore, the study of structure, phase composition after mechanical treatment of multicomponent mixtures is actual question.

In this work it has been studied the effect of mechanical treatment of Ti-Ni at near-equiatomic composition of powder on structure and phase composition.

2. Material and methods

Near equiatomic $Ti_{0.45}Ni_{0.55}$ powder (produced by the «Polema» plant) with average particle size of 11.1 μm with 7.5 μm of standard deviation is used for studies.

The mechanical treatment was performed in an AGO-2 high-energy planetary ball mill using the container and stainless steel balls with 6…10 mm diameter. The powder mixture was treated at a rotation speed of 1820 rpm; acceleration during treatments was 60g. The ratio of the mass of balls and powder was 5:1. The mechanical treatment was carried out for different time 5 – 300 second.

The morphologies of powders were observed by scanning electron microscopy (SEM) TESCAN VEGA 3SBH. The measurements of particle size distribution of powders by means of laser diffraction (LD) method were performed with SALD-7101.

Analysis of the phase and structural transformations was carried out with diffractometers with CuK_α radiation using step-by-step registration and counting time for obtaining statistical accuracy better 3 %. XRD line profile analysis was used to determine the coherently diffracting domain (CDD), which calculated by the Scherrer equation [8] for the first spectra line. The lattice parameters were determined using an application package for X-ray analysis.

The powders specific surface areas (SSA) were measured with a Sorbi™ series instrument measurement device. From the obtained values of the specific surface, it has been calculated the average particle size assuming their spheroids shapes.

3. Results

SEM images showing morphology of powders after mechanical treatment after 30 s (a); 300 s (b) are shown in Fig. 1. One can see it decreases of size particles from 11.1 to 7.8 μm after 30 s of mechanical treatment due to the fracture of particles, and then it increases of size particles to 33.2 μm after mechanical treatment 300 s due to agglomeration. The form of particles changes from angular to spherical one. These results correspond to measured powder apparent density, when it increases from 1.96 to 2.56 g/cm^3 and then decreases to 2.38 g/cm^3 after 30 and 300 s mechanical treatment, respectively.

(a) (b)

Figure 1. SEM images showing morphology of powders after mechanical treatment: 30 s (a) and 300 s (b)

In Fig. 2 shows the dependence of the average particle size measurement different methods and CDD at different time of mechanical treatment. The average particle size measured by method laser diffraction decreases from 8.0 to 0.5 μm with increasing of mechanical treatment time. The average particle size calculated from the obtained values of the specific surface do not significantly changes 3.9 μm – 1.7 μm. The CDD size decreases from 43 nm to 5 nm after mechanical treatment, that is significantly less than the average particle size measured by methods SEM, LD and BET. That is to say, particles consist of several crystallites, so number of crystallites increases from 5×10^4 to 13×10^6 with increasing time of mechanical treatment.

In Fig. 3 shows XRD patterns of TiNi powders before and after mechanical treatment. As one can see the powder before mechanical treatment consists of a mixture of phases such as TiNi (B2 and 19\`), $Ti_2Ni$ ($E9_3$) and $Ni_3Ti$, vertical lines are indicated them the angular positions. The lattice parameters calculated from diffraction lines are in good agreement with the literature data [9-12], TiNi (B2) the parameter is 0.30074 nm, TiNi (B19\`): a = 0.26902 nm, b = 0.45698 nm, c = 0.41215 nm, β = 87.412 °; Ti_2Ni a=1.13332 nm, $TiNi_3$ a = 0.50989 nm, c = 0.83077 nm. From the measured angular positions of the lines with increase mechanical treatment time it has been calculated lattice parameters which decreases by 0.3 % for TiNi, $TiNi_3$ parameters decrease not more by 0.8 %, for Ti_2Ni this change was 0.5 %.

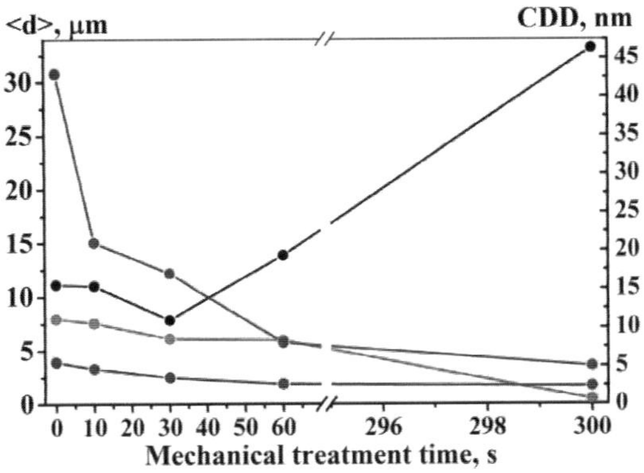

Figure 2. The dependence of the average particle size measurement different methods and CDD on the time of mechanical treatment: —•— - BET, —•— - LD, —•— - SEM, —•— - CDD

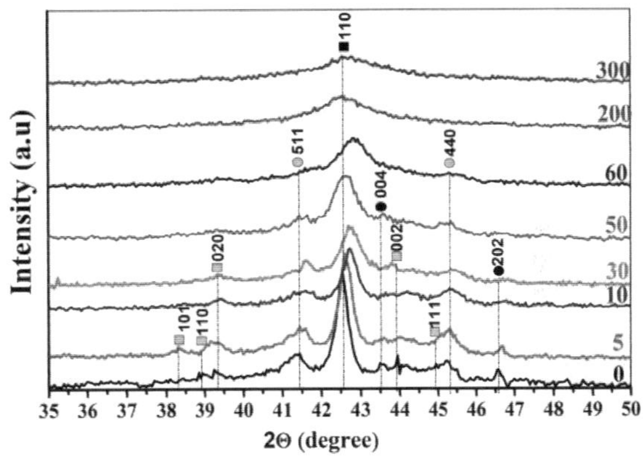

Figure 3. XRD patterns of TiNi powders before and after mechanical treatment (5 – 300 second): ■ – TiNi (B2); ■ – TiNi (B19\`); • – Ti_2Ni ($E9_3$); • – Ni_3Ti ($D0_{24}$)

The dependence of the width of peaks with mechanical treatment time is shown in Fig. 4. As one can see, there are two stages during treatment: on the first step up to 10 s the width of all peaks doesn\`t significantly change, but on the second step after 10 s the width of all peaks increases for mechanical treatment, it can be connected to an increase in the defects density in the crystal lattice which according to literature data [13].

As one can see from Fig.5 after mechanical treatment quasi-amorphous phases during treatments appeares on x-ray patterns and its content increase for an increasing of milling time. This result is shown in inset in Fig.5, the arrows indicate X-ray lines that correspond to the quasi-amorphous phase. The relative content of the quasi-amorphous phase increases to 55 % with increasing mechanical treatment time, while the content of the crystalline phases decreases.

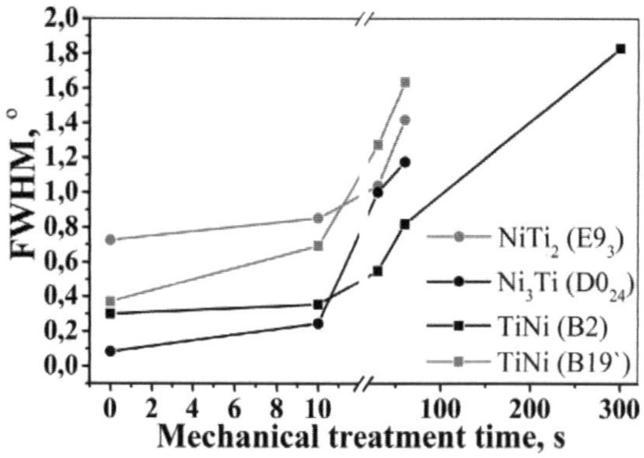

Figure 4. The dependence of the width (FWHM) of peaks on mechanical treatment time

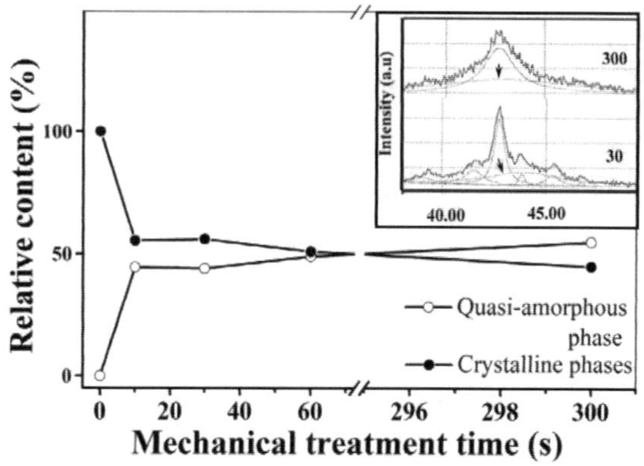

Figure 5. Relative content of crystalline and quasi-amorphous phases at different times of mechanical treatment. The arrows indicate X-ray lines that correspond to the quasi-amorphous phase

The grinding process can evaluate using the equation according to [14], characterizing the kinetics of the fine-milling process (Eq. (1)):

$$R = R_0 e^{-k\tau^m} \quad (1)$$

where R is particles content in coarse fraction after milling for time τ, R_0 is particles content in coarse fraction in the initial material, k – kinetic coefficient of the milling; m – grinding mechanism constant. In [14] it has been shown that more brittle materials have higher grinding rates k/m.

Therefore, from the dependence of the average volume particle size on the time of mechanical treatment (Fig. 6.) can calculate k and m. It was shown that k is 3.22, m is 0.05, and so k is more than m by 64 times, that mean the particles are brittle and quickly grinding by balls with an increase mechanical treatment time. However, on the SEM results (Fig. 1, a and b) can see that form of particles change that mean he plasticity of the powder. Thus, two processes compete during mechanical treatment, characterizing either brittlness or plasticity. Many defects appear during mechanical treatment stipulates a quasi-amorphous state with increase of mechanical treatment time, that can be seen in Fig. 5.

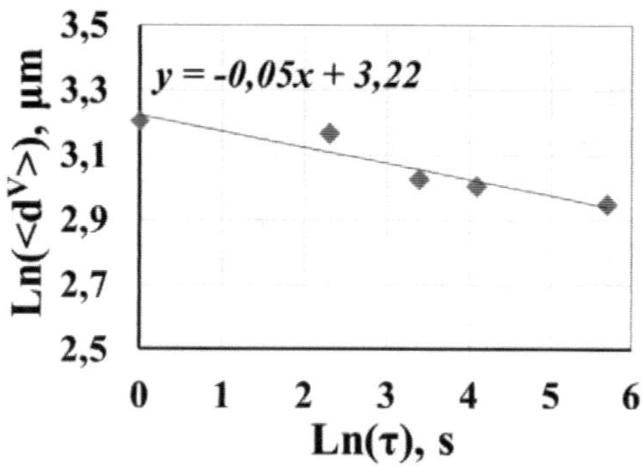

Figure 6. The dependence of the volume average particle size of powder calculated using volume distribution particles on the time of mechanical treatment

4. Conclusion

It was shown that the average particle size measured by scanning electron microscopy changes not-monotonous after mechanical treatment due to destruction and subsequent agglomeration particles. A state with a very small coherently diffracting domain size (5 nm) is formed after mechanical treatment 300 s, which corresponds to quasi-amorphous state of material. The relative content of quasi-amorphous phase increases to 55 %. The coherently diffracting domain size is significantly less than average particle size measured by laser diffraction and the average particle size calculated from the obtained values of the specific surface. Therefore, the powder particles consist of several crystallites, the number of which increases from 5×10^4 to 13×10^6 with mechanical treatment time. Two stages during treatment can be detected, on the first step up to 10 s the widths does not change, on the second step up after 10 s the widths changes in result of increasing the defect density. The lattices parameters of all phases decrease after mechanical treatment, in particular, the lattice parameter TiNi decreases by 0.3 %, $TiNi_3$ parameters decrease by not more than 0.8 %, for Ti_2Ni this change is 0.5 %.

Thus, one can identify two processes competing, characterizing either by fragility (changing the average particle size) or by plasticity (changing geometric form of particle and formation quasi-amorphous phase) during mechanical treatment.

Acknowledgement
The reported study was funded by RFBR, project number 19-38-90196.

References
[1] Skakov Yu A 2004 *MiTOM* **4**, 3–12.
[2] Zadorozhnyi V Yu, Milovzorov G S., Skakov Yu A, 2008 *MiTOM* **8**, 46–52.
[3] Schwarz R B, Koch C C1986 *Appl. Phys. Lett.* **49(3)**, 146–148.
[4] Boldyrev V V. 1986 *Proc. Indian nath.Sci.* Acad. **52**, 400–417.
[5] Smetkin A A, Pimenova N V et. al. 2004 *Poroshkovaya metallurgiya* **27**, 61–64.
[6] Otsuka K, Ren X 2005 *Prog. Mater Sci.* **50**, 511–678.
[7] Tang W, Sundman B, Sandström R et. al. 1999 *Acta Metall.* **47**, 3457–3468.
[8] Scherrer P. 1918 Göttinger Nachrichten Gesellschaft. 2 98–101.
[9] Stuewe H-P, Shimomura Y Z 1960 *Metallkd* **51**, 180.
[10] Sinclair M G 1981 *Acta Crystallogr. Sec. E* **37**, 1803.
[11] Laves F, Wallbaum H J 1939 *Z. Kristallogr* **101**, 78.
[12] Mueller M H, Knott H W 1963 *Trans. Met. Soc. AIME* **227**, 674.
[13] Gattia D M, Gizer G, Montone A 2014 *Int. J. Hydrog. Energy.* **39**, 9924–9930.
[14] Tadzhiev F Kh, Ismatova R I, Shamuratova Sh M 1991 Glass Ceram. 48(6), 255–257.

IOP Conf. Series: Materials Science and Engineering **696** (2019) 012019 doi:10.1088/1757-899X/696/1/012019

Kola Peninsula in solving problems of national arctic materials science

A I Nikolaev and S V Krivovichev

Federal Research Centre "Kola Science Centre of the Russian Academy of Sciences" Apatity 14 Fersman street Russia

E-mail: nikol_ai@chemy.kolasc.net.ru, skrivovi@mail.ru

Abstract. The Kola Peninsula is one of the most developed Arctic regions in Russia and in the world. Its share in the Russian production of apatite, nepheline, loparite and baddeleyite concentrates is 100%. The region produces up to 45% of nickel, and 13% of copper, and the majority of rare-earth metals, niobium, and titanium compounds from the regional minerals. The ores mined from the existing deposits can be used for production of up to 30 various concentrates. The materials produced from the regional minerals include metal nickel, cobalt, copper, titanium, niobium, rare-earth metals, steel; sealants, components of electrode coatings and welding fluxes, sorbents for disposal of radioactive waste and wastewater containing heavy non-ferrous metals and other contaminants, high-porous heat insulation and fire-resistant materials for arctic application, cement, concretes with improved resistance to frost and sea water. In satisfying demand for new materials, the first and foremost objective is to establish a pilot engineering base for adaptation of process flowsheets, and collection of data for design and construction of new operations. Successive implementation of this objective requires concentrated efforts of scientists and engineers focused on implementation of the Kola Chemical and Technological Cluster Project, which enables to improve significantly the economy of minerals use by means of products of their deep processing. At the first stage of the Cluster operation, a process of new titanium sorbents production will be established to solve environmental issues — treatment of non-ferrous metals industry wastewater from heavy metals, and processing of liquid radioactive waste accumulated in the Arctic zone of Russian Federation.

1. Introduction

The Arctic zone of the Russian Federation (hereinafter — AZRF) is the largest zone in the world arctic sector. It is characterized by high density of mineral deposits and the importance of its mineral resource base for national economy. The AZRF accommodates about 10% of active world nickel resources, about 19% of platinum group metals, 10% of titanium, and over 3% of zinc, cobalt, gold and silver, as well as the largest amount of rare-earth metals (REM) [1]. There are quite a number of problems related to mining and rational use of the unique arctic mineral resources, their processing and production of special materials for development of the Northern areas in Russia, or production of such materials from arctic minerals in other regions of the country. Actual situation in arctic materials production needs higher level of research and development in the field of arctic material science, modern research and production equipment, competitive capacity of the products, demand in materials produced (partially, and due to lack of consumers' money) [2–4].

At the same time, development of vast northern territories is impossible without materials that can maintain vehicles performance under extreme temperatures, and ensure appropriate living conditions for people at these areas. The Kola Peninsula minerals processing waste can be a source for such materials production [5].

2. Murmansk region as an important part of the Russian resource base for development of arctic material science

The Kola Peninsula is an important supplier of structural and strategic materials to national and foreign markets. Without stable supply of such materials to the national industry, it is impossible to guarantee the national safety. For example, JSC Kolskaya Mining and Metallurgical Company produces up to 45% of metal nickel, and other metals. However, the main products of the region are ores and mineral concentrates for production of functional materials used in various industries. The Murmansk region economy is mainly focused on production of mineral concentrates as final products. The examples include regional mining and processing operations: JSC Apatit, Kovdorskiy GOK, Olenegorskiy GOK, North-Western Phosphorous Company, Lovozerskiy GOK. The Kola Peninsula is an important Russian base for various mineral resources [6, 7].

The most important federal resources are apatite-nepheline ores (ANO) of Khibiny deposits, loparite ores of Lovozerskie deposits, perovskite-titanium magnetite ores of Afrikandskoe and Vuoriyarvinskoe deposits, ilmenite ores of Gremyakha-Vyrmess deposits, magnetite ores in Kovdorskiy and Olenegorskiy districts, kyanitic ores in Keytvy, deposits of platinum and non-ferrous metals, and a lot of other deposits [6, 7]. The Kola Peninsula host significant portion of Russian apatite, nepheline, kyanite, titanium, rare-earth, vermiculite, phlogopite, and other resources. The resources mined in the region can used for production of up to 30 various concentrates, with can satisfy both the demands of AZRF, and the whole country.

Only at the Khibiny deposits, during 90 years of their operation, 2 billion tons of apatite-nepheline ore were mined. In the Soviet Union, and now in Russia, this was the main mineral resource for production of phosphorous and its compounds. Such ore contains also other valuable rock-forming titanium minerals — sphene, and titanium magnetite, which are not recovered now, and disposed to tailings storage facilities [8].

The first studies on the ANO use were conducted by academician A. E. Fersman in 1932. The studies were focused on recovery of all valuable components from complex ANO and ores from the nearest deposits, and satisfying the national industry demands in deficient products. The list of the main products included phosphate fertilizers, thermophosphates, yellow phosphorus, ferro-phosphorus, phosphoric and sulfuric acids, liquid glass, alumina, cement, concentrate of rare metals, non-ferrous metals, etc. Moreover, the planned production scale of some products shall reach from thousand tons per year up to million tons per year.

3. Scientific basis for the Kola resources processing. Development and improvement of process flowsheets

Many papers and books were written on mineral resources of the Murmansk region, including results of scientific and production companies' studies, exploration programs. They enable to understand availability of minerals, their reserves, and existing technical solutions on processing [6–9].

Significant portion of the Kola Peninsula is associated with non-traditional resources. There was almost no experience in processing of such resources in the world industry. The Federal Research Center "Kola Scientific Center of the Russian Academy of Sciences" (FRC KSC RAS), in cooperation with Russian industrial and academic institutions, universities and existing operations processing resources mined in the Murmansk region, conducts systematic studies of the regional resource base, ore mining, development of processing methods, processing of concentrates and production of materials. The FRC KSC RAS includes nine scientific institutions in the form of separate subdivisions, including geological, mining, chemical, environmental, economical, and energy institutions. Close cooperation between them enables to maintain high potential of operations, and creates opportunities for deeper processing, and obtaining of products with better consumer-oriented characteristics. The Rare-Earth Metals Chemistry and Technology Institute of the FRC KSC RAS developed the basic technology package of non-traditional minerals processing [6, 9]. It enables to choose the most sustainable resource-saving technology options providing synthesis of a wide range of products, which is important under unstable conditions at the consumer markets of titanium, rare metals and their compounds.

In recent years, the FRC KSC RAS has developed technologies for synthesis of some rare titanosilicates minerals analogues as promising polyfunctional materials, for example, molecular sieve, titanosilicate "nanopuzzles", cation exchangers, photocatalyst, etc.. Their composition, morphology, and properties are characterized. The synthetic analogs of the ivanyukite-group titanosilicates (SIV) are obtained [10]. A principal scheme of titanite concentrate processing for SIV synthesis are given in Figure 1

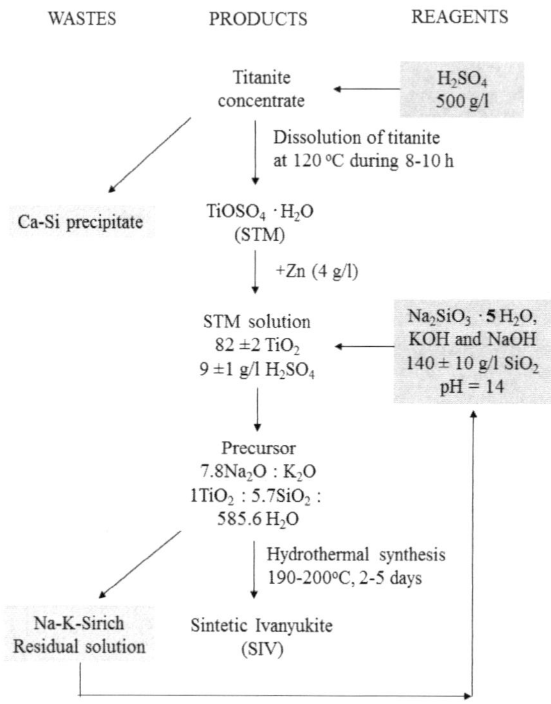

Figure 1. A principal scheme of titanite concentrate processing for SIV synthesis.

The framework composes corner-sharing SiO_4 tetrahedra and TiO_6 octahedra linked through bridging oxygen atoms and provide pore system with channels containing H_2O molecules and exchangeable Na^+ cations. It has been found that SIV has high selectivity towards Cs^+ and Sr^{2+}. The direct calcination of the ^{137}Cs-^{90}Sr-loaded SIV leads to the formation of the Synroc-type titanate ceramics. The synthetic ivanukite with wide-pore microporous framework is considered to be perspective cation-exchanger for the treatment of non-ferrous metals industry wastewater from heavy metals, and processing of liquid radioactive waste accumulated in the Arctic zone of Russian Federation.

The scientific basis for processing of complex resources of the Kola Peninsula has been studied in sufficient detail by now. The developed technologies demonstrate the ability to recover almost all valuable components from mineral concentrates; however, in practice, the efficiency of using mineral resources is not high enough, which is explained by low market demand, as well as low economic indicators of some potential products recovery.

Rare-earth metals registered as national resources in the Khibiny ores are among the "lost" components of apatite concentrate processing. Over 100 thousand tons as recalculated to REM oxides are lost annually during production of apatite concentrate. 10% of this amount are presented by the most deficient REM of medium-weight group. At the same time, the technology of collective REM concentrate recovery from apatite and phosphogypsum was tested at pilot scale in the operations of JSC PhosAgro, Cherepovets, JSC Acron, Velikiy Novgorod, GK Yuniskhim, LLC, Voskresensk, and the technology of individual REM recovery was tested at large laboratory scale in NPK Rusredmet, LLC [11, 12]. The prospects of such technologies implementation are determined by actual demands of the national industry. Establishment of enterprises based on the principle of public-private partnership could accelerate implementation of these technologies in case of any government orders for produced rare-earth metals, including those for state reserves of strategic materials.

Promising products for development of the Arctic regions are components of electrode coatings and welding fluxes based on mineral concentrates from ores of the Kola Peninsula deposits — sphenic, nepheline, titanomagnetite, quartzite, dolomite, olivinite, and other concentrates [13]. Such materials, which ensure higher welding and processing properties, in particular, higher cold resistance of welded steels, have undergone extensive tests at JSC Sevmash, and a number of welding operations under the patronage of CRISM "Prometey". Lower demand of the welding industry in mineral concentrates (from hundreds of tons to thousands of tons per year) is an obstacle to production of non-core low-tonnage materials at large mining and processing operations. At the same time, the use of imported welding materials is growing, and their share currently exceeds 50%. Without the state support (government order, preferences for companies producing import-substituting products, including for the Arctic region), we cannot change this situation.

The ideas stated by academician A. E. Fersman in 1932, related to sustainable use of mineral resources in the Murmansk region, were noticeably ahead of their time, and needed improvement. By present, the concept of the Kola Chemical and Technological Cluster Project (KCTC) has been developed in detail at FRC KSC RAS. The cluster is focused on production of high-tech materials for the national industry. Basically, such materials are low-tonnage. In the 1980s, in Apatity, a large plant for production of lithium niobate was built for the civil sector of the economy and the military-industrial complex of the Russian Federation. A small enterprise for production of high-capacity tantalum condensate powders was established. JSC Apatit, with participation of KSC RAS, started to produce coagulants-flocculants, explosives from products of nepheline processing, components of special-purpose welding materials, and other products. The importance of these works for the country was recognized with the RF Government Award in 1997, and the RF National Award in 2000.

4. Conclusion

Depletion of reserves of many operating deposits in Russia and other countries increases attractiveness of a new resource base of the Arctic zone, including mineral resources of the Kola Peninsula, and necessity for its use in the near future.

The developed fundamentals of chemical technology for the Kola Peninsula complex resources, and efficient processing flowsheets with production of materials for modern advanced industries enable to satisfy the demand in structural and strategic materials (non-ferrous, rare-earth, noble, ferrous metals, their compounds, materials for electronic equipment, catalysis, sorption, medicine and other purposes) not only in the Arctic zone, but also in the whole country. Production of materials for construction industry on the basis of overburden, ores, mineral concentrates, beneficiation and chemical processing waste can and should lead to an increase in the efficiency of resources use, a noticeable decrease in the amount of industrial waste, and improvement in the environmental situation.

Establishment of a pilot industrial base for adaptation of processing flowsheets, production of high-tech product batches, and collection of data for design and construction of industrial operations becomes an important priority objective in satisfying demand for new materials. The success of this objective is determined by concentrated efforts of scientists and regional industrial operations focused on implementation the KCTC project, which increases completeness and economic indicators of resources use. At the initial stage of the KCTC project, a pilot production of new titanium-containing sorbents will be established to solve environmental problems — treat non-ferrous metallurgy from heavy metals, and process liquid radioactive waste accumulated in the Arctic zone of Russia. The KCTC establishment will significantly accelerate implementation of innovations, improve the economic security of the country, and ensure the production of modern competitive products. In particular, it is planned to create a production of sealants for the aerospace industry, tantalum and niobium condensate powders, materials for acoustic-optoelectronics at national scale. It is important to note that further development of the mining industry and solution of socio-economic problems of the region can be carried out without any noticeable increase in the environmental impact and environmental degradation.

Acknowledgement
The reported study was financially supported by RFBR according to the research project 18-29-121039 and was partly supported by the theme of state assignment of AAAA-A17-117020110035-5

References

[1] N P Laverov, V I Vasilyev and A A Makosko 2015 Scientific and technical problems of the Arctic development. Scientific session of the General Meeting of the Russian Academy of Sciences, December 16, 2014. (Moscow: Nauka) 490 p. [In Russian]
[2] Buznik V M and Kablov E N 2017 Current Situation and Prospects of Arctic Materials Science. Bulletin of the Russian Academy of Sciences, **87** **9** 827 [In Russian]
[3] Buznik V M, Burkovskaya N P, Zibareva I P, Cherepanin R N 2017 On the issue of a roadmap development for the Russian Arctic materials science. Materialovedenie [Materials Science], **4** 8 [In Russian]
[4] Buznik V M, Burkovskaya N P, Zibareva I P, Cherepanin R N 2017 On the issue of a roadmap development for the Russian Arctic materials science. Materialovedenie [Materials Science], **5** 22 [In Russian]
[5] Construction and technical materials produced from mineral resources of the Kola Peninsula 2003. Makarov V N, Krasheninnikov O N, Gurevich B I et al. Part 1, 2 (Apatity: KSC RAS publishing house) 430 p. [In Russian]
[6] Kalinnikov V T and Nikolaev A I 2011 Prospects for the development of chemical industry raw materials based on the Kola peninsula *Tsetnye metally* **11** 17 [In Russian]
[7] Fedorov S G , Nikolaev A I , Brylykov Yu E et al. 2004 Chemical processing of mineral concentrates of the Kola Peninsula (Apatity: K&M publishing house) 198 pp. In Russian.
[8] Nikolaev A I, and Krivovichev S V 2019 Prospects for the Development of the Kola Chemical Technological Cluster in Transition from a Resource-Based Economy to an Innovative Economy, *Theoretical Foundations of Chemical Engineering* **53** **5** 980
[10] Nikolaev A I and Krivovichev S V 2017 Natural minerals and their synthetic analogues as prototypes of functional materials (based on the experience of the Kola Nanomaterials Research Centre) *Bulletin Tomsk State University. Chemistry* **8** 7 [In Russian]
[11] Lokshin E P and Tareeva O A Technologies development for the extraction of rare earth elements while sulfuric acid processing of the Khibiny apatite concentrate for mineral fertilizers production 2015 (Apatity: KSC RAS publishing house) 268 p. [In Russian]
[12] Polyakov E G, Nechaev A V and Smirnov A V 2018 Metallurgy of rare metals (Moscow: Metallurgizdat) 732 p.
[13] Nikolaev A I, Brusnitsin Yu D, Vasilieva N Ya et al. 2000 Production of Welding Materials from Minerals of North-West Russia Mineral Processing and Extractive Metallurgy Review **22** 1-3 p 273.

IOP Conf. Series: Materials Science and Engineering **696** (2019) 012020 doi:10.1088/1757-899X/696/1/012020

Investigations of the stress-strain state of a compressed steel concrete column using digital image correlation method

A S Plyaskin[1], A A Klopotov[1], A I Potekaev[2] and B O Koshko[1],
A M Ustinov[1], O G Volokitin[1], I.A. Kurzina[2] and G I Tayukin[1]
[1]Tomsk State University of Architecture and Building, Tomsk, Russia
[2]National Research Tomsk State University, Tomsk, Russia

E-mail: plyaskinandrei@mail.ru

Abstract. The results of the investigation of deformation during axial compression of a steel-concrete round column with dimensions of 1000×102 mm on the testing machine Universal Testing Machine 4500 using a digital optical system Vic-3D are presented. The patterns of the evolution of strain field distributions on the surface of the steel shell of the column are obtained. A deformation diagram indicating the transition stages of elastic strains to inelastic ones is constructed. A jump-like dependence of the Poisson's ratio on the longitudinal relative strain of the steel concrete column during deformation of the steel concrete column is established. This phenomenon is explained under the assumption that the deformation process involves redistribution of the forces (deformations) among the components of the composite system at the stage of joint operation before separation of the steel/concrete media and operation of concrete beyond elastic strain under conditions of volumetric compression.

1. Introduction

The use of complex structures perceiving compression, such as steel-concrete and those consisting of a jointly operating steel shell of a concrete core, has several advantages as compared to reinforced concrete structures [1-3]. The strength of a concrete body pressed by a steel shell exceeds the strength of 2 times. A positive effect of increasing strength characteristics of the concrete core is observed under both static and dynamic loads. As compared to conventional reinforced concrete structures, the use of steel-concrete allows reducing the material consumption of steel-concrete in 1.5–2 times, and, therefore, the mass of structures [4,5]. These and other positive qualities allow the effective use of steel-concrete structures as columns of high-rise buildings [6,7], as well as the use in seismic regions and regions of the far north. The behavior of the "shell-core" system operating under conditions of the volumetric compression is of great interest to researchers in this field. The study of the distribution of surface deformation of the steel shell and the deformation curve of the entire sample can clarify the redistribution of forces among the components of the system. One method for analyzing deformation of a steel-concrete column at mesoscale and macroscale levels is the digital image correlation method.

Despite numerous studies conducted by domestic and foreign authors, steel-concrete structures are not widely used in Russia. One of the reasons is the lack of regulatory documents and calculation methods that allow normalizing the limiting state of a steel-concrete structure. Verification of the calculation model with the results of experimental studies will improve the accuracy of calculations and ensure the reliability.

The purpose of the paper is to identify features during compression strain of a steel-concrete column based on experimental studies of the distributions of relative longitudinal and transverse straining the near-surface layers of a steel pipe.

2. Materials and methods

Steel-concrete columns were used as prototypes for studying the stress-strain state of the "shell-core" system. The clip was made of a steel pipe with a height of 1000 mm, a diameter of 102 mm, and a wall thickness of 3 mm. The concrete core filling the inner space of the shell was made of monolithic fine-grained heavy concrete. The proportion of the components of the concrete core was taken in the ratio by weight of 2.35:1:5.5 cement: sand: screening. The screening with a fraction size of 5 mm was taken as a filler. The accepted water-cement ratio was W/C=0.46. The prismatic strength of control concrete samples at the time of testing was 27.1 MGa.

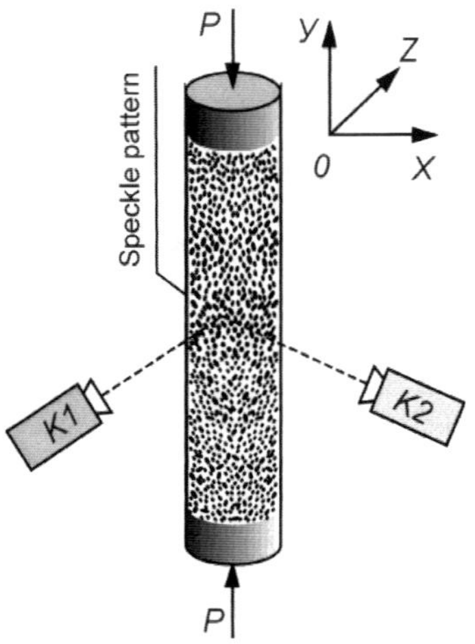

Figure1. Scheme for recording images from a speckle pattern on the surface of a steel-concrete column under compression: *K1* and *K2* are digital cameras; *P* is the applied load

Thetestsamplesofsteelconcretecolumnsweresubjectedtostaticaxialcompression. The columns were deformed using the UTM-4500 hydraulic unit at a speed of 2 mm/s. The test program provided for stepwise loading of columns. A predetermined level of compressive force was held for 5 minutes in order to redistribute stresses along the sample surface. Each stage of the test corresponded to an increase in the compressive strength of 60 kN. The Vic-3D stereoscopic measuring system was used to register the evolution of relative and absolute strain of steel-concrete columns, [8, 9]. Figure1 shows the test scheme. Synchronous fixing of speckle-patterns from the surface of the columns was carried out using two cameras and the VicSnap software.

The operation principle of the Vic3D digital optical system is correlation of digital stereoscopic images. The measurement results have allowed obtaining data on the displacement of microvolumes in a three-dimensional space. The resulting patterns of displacement fields are projections of the displacements of local surface areas along the OX axis "transverse strain", along the OY axis "longitudinal strain", and along the OZ axis "transverse strain" (figure1). The deformation pattern of the steel shell surface is obtained as a result of combining changes in microregions. The speckle-structure in the form of randomly distributed points on the sample surface was ensured by spraying a contrast finely dispersed aerosol paint.

3. Results and discussion

Figure 2 presents the deformation curves in $\sigma = f(\varepsilon_{YY})$) and $\sigma = f(\varepsilon_{XX})$ obtained under static compression of the samples of steel-concrete columns. It is seen that the elastic section is present on the deformation curves (AB sections in figure2). Then, an increase in the applied load leads to plastic strain of the column (sections *BC* in figure2). Further deformation of the column is accompanied by localization of strain fields in the central part of the column indicating a loss of stability due to longitudinal bending. This is very clearly seen in the three-dimensional distribution of longitudinal relative strain of the steel-concrete column surface (figure3).

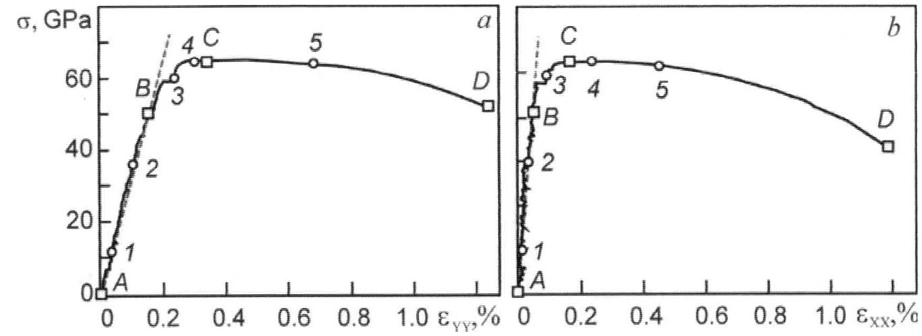

Figure 2: Deformation diagram of the steel-concrete column in the longitudinal (a) and transverse (b) directions. The numbers correspond to the position on the curve of the patterns of deformation structures (figure 3 and 4). AB – is the elastic strain stage; BC is the non-elastic strain stage

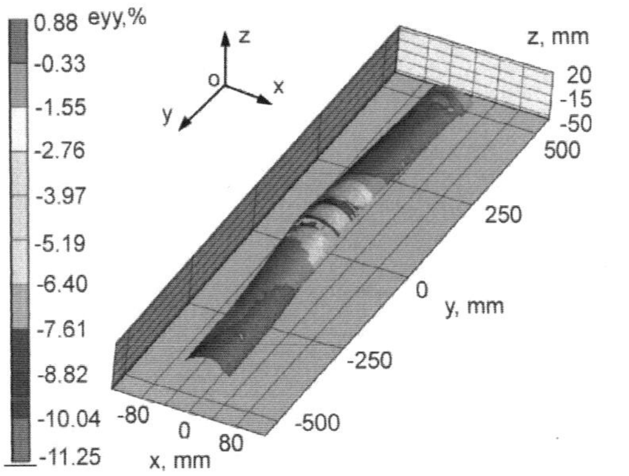

Figure 3. Three-dimensional distribution of vertical relative deformations on the surface of the steel-concrete strut under a stress of 63 GPa (point 5 figure 2)

The results of experimental studies of the axial compression of the column have allowed obtaining distribution patterns of strain fields on the surface of the steel-concrete column in the longitudinal (figure 4) and transverse (figure 5) directions. On the presented patterns of longitudinal distributions of strain fields on the surface of the column, it is seen that distribution patterns of the deformation along the surface of the column are heterogeneous from the beginning of loading until the moment of the stability loss. An increase in the load leads to redistribution of strain fields and formation of local strain foci.

In the elastic stage of compression, the density of local strain foci in the supporting zones of the column is higher than in the middle part of the column. Local foci with positive and negative values of longitudinal strains are formed in the lower part of the column (section AB, points 1 and 2 in figure 2). Whereas strain foci with negative values are formed in the upper part of the column from the very beginning of compression. Such a noticeable heterogeneity in distribution of strain foci is not observed in the patterns of transverse strain fields (patterns 1 and 2 in figure 5).

Transition from the elastic stage to the region of inelastic strain of the column (figure 2, section of the BC curve, points 3 and 4) is reflected in the distribution patterns of strain fields in the appearance of local plastic strain foci on the surface of the column in its central part (figure 4 and 5). This is clearly seen in the schematically presented distribution patterns of longitudinal strain fields in the upper part of the column (figure 6). A decrease in the regions (regions with ε_6) with highest strain values takes place in the region of the upper punch of the testing machine on the surface of the steel shell (figure 4, 5, regions with ε_6 figure 6). This indicates that stresses are transferred from contact areas of the column with the punches of the testing machine to the central region of the column. Moreover, distribution patterns of strain fields show that this stage includes a complete disappearance of local strain foci with positive values in upper and lower parts of the column during deformation in the longitudinal direction.

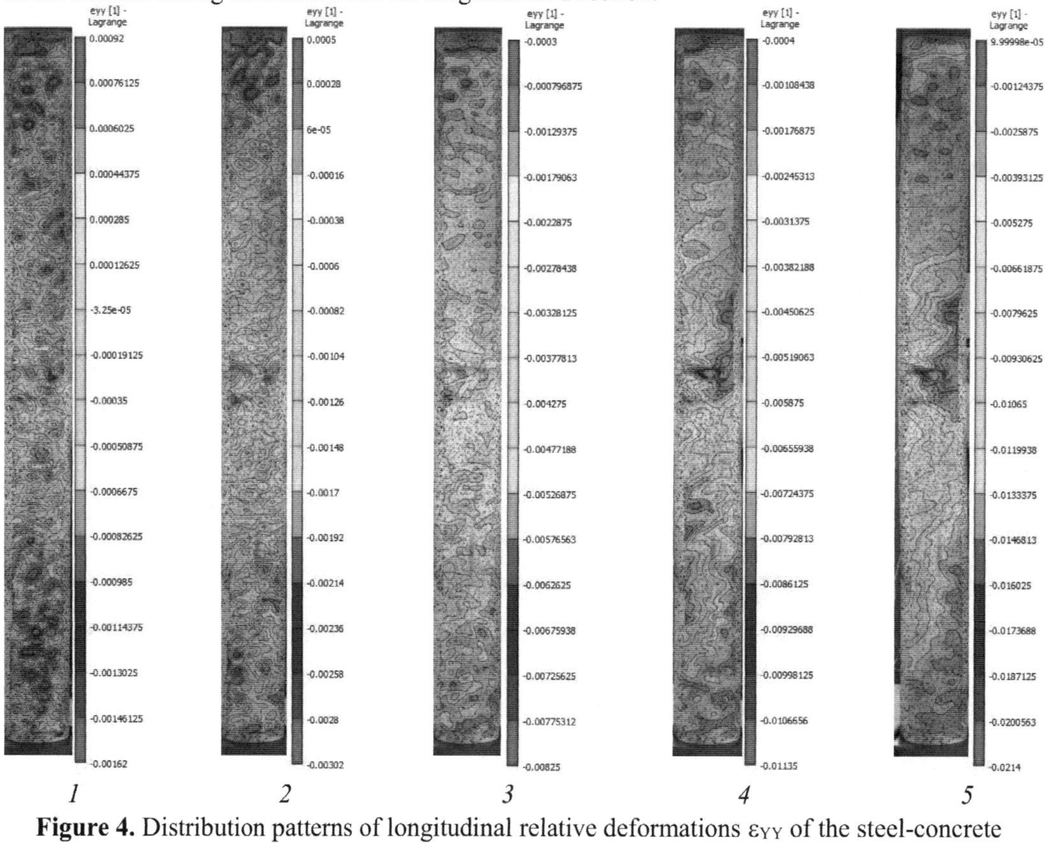

Figure 4. Distribution patterns of longitudinal relative deformations ε_{YY} of the steel-concrete column surface. The numbers on the diagram in figure 2 show the corresponding strain-strain states of the column during the test

Fusion of local strain foci into vast areas is observed in upper and lower parts of the column in the patterns of transverse and longitudinal strain fields. Local fields with lower deformation values are formed inside these vast areas (figure4, 6). Local areas with deformation ε_1 are shown in figure 6. The size and concentration of these areas increase with increasing external stress (pattern 3–5 figure 6). The appearance of local areas with lower deformations on the surface of the steel shell is associated with redistribution of stresses in the steel shell due to its expansion by the concrete core [6].

An asymmetric distribution of plastic strain foci relative to the longitudinal axial line of the column takes place in the central part of the column. Subsequently, an increase in the applied external load leads to a longitudinal bending of the column in the central part. This phenomenon indicates a loss of stability of the column.

The Poisson's ratio μ plays an important role in studying the deformation effect on solids when their linear and/or volumetric dimensions are changed. This parameter characterizes the property of a solid to retain its initial volume at the elastic stage of deformation. Calculation of the Poisson's ratio under uniaxial tension is determined by the ratio of the transverse strain of the sample $\varepsilon_{XX}=\Delta h/h_o$ to its longitudinal elongation $\varepsilon_{YY}=\Delta l/l_o$ according to the formula [10]:

$$\mu = \frac{\varepsilon_{XX}}{\varepsilon_{YY}}. \qquad (1)$$

Where h_o is the initial width of the sample, Δh is the change in the width of the sample during deformation, l_o is the initial length of the sample, and Δl is the change in the length of the sample during deformation.

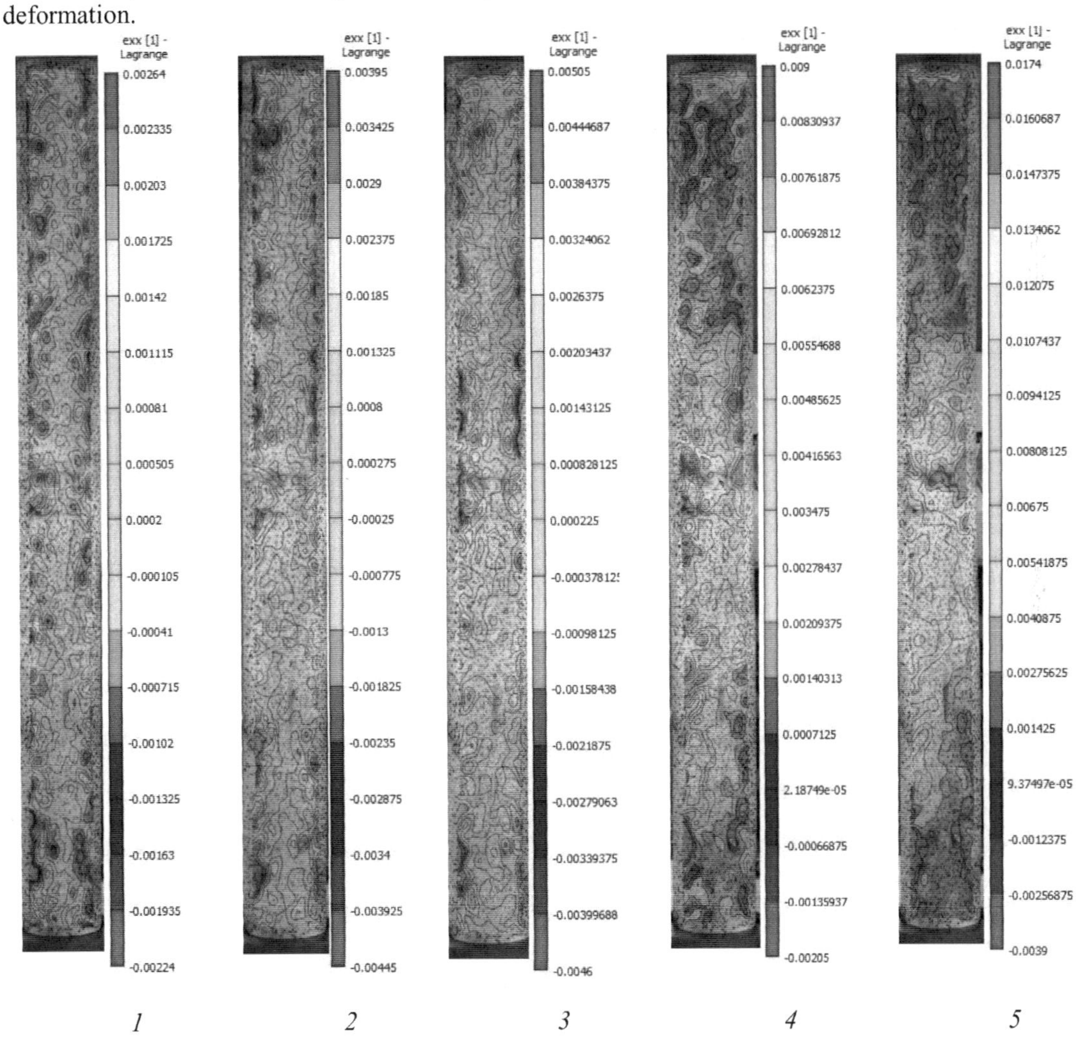

1 *2* *3* *4* *5*

IOP Conf. Series: Materials Science and Engineering **696** (2019) 012020 doi:10.1088/1757-899X/696/1/012020

Figure 5. Distribution patterns of transverse relative deformations $\varepsilon_{XX}$of the steel-concrete column surface. The numbers in the diagram in figure 2 show the corresponding stress-strain states of the column during the test

The dependence of the Poisson's ratio on the longitudinal relative strain ε_{YY} on the surface of the steel-concrete column has been obtained using the above mentioned dependences in figure 2 using the formula (1) (figure 7). The jump-like nature of deformation (section AB, figure 7) is very well pronounced in the dependence $\mu=f(\varepsilon_{YY})$ at the initial stage of deformation of the steel-concrete strut in the elastic region under low external applied stresses. The jump-like behavior of the deformation ends at the *BC* stage of inelastic strain. Then, deformation of the steel-concrete column in the *CD* section is accompanied by a longitudinal bending of the column. At this stage, deformation of the column occurs smoothly. This nature of the jump-like reaction of the steel-concrete column to the deformation effect is conditioned upon its compositional structure: a concrete core isolated by metal sheathing. Compression strain of such a composite material with different values of the Poisson's ratio of the core and shell materials leads to failure of the adhesive bond at the concrete/steel interface, and manifests itself in the form of ajump-like nature of deformation curves.

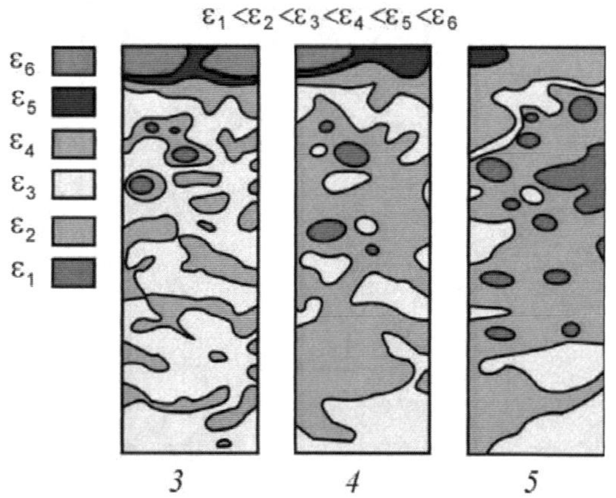

Figure 6. Schematic representation of fragments of distribution patterns of vertical relative deformations on the surface of the steel concrete column at stages *BC* and *CD*. The numbers below the fragments of the patterns correspond to the points on the deformation diagram in figure 2. .ε_1, . . . , ε_6 correspond to deformations in various regions on the steel-concrete column surface

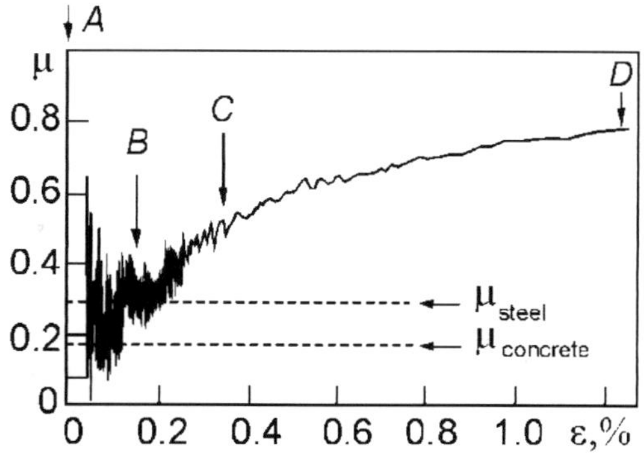

Figure 7. The dependence of the Poisson's ratio on the longitudinal relative deformation ε_{YY} on the steel concrete column surface. Sections *AB*, *BC*, and *CD* correspond to stages in the deformation dependence in figure 2

The Poisson's ratio in the elastic strain region on average coincides with the Poisson's ratio for steel (figure 7). Upon transition to the *BC* stage, the value of the Poisson's ration begins to increase, and then at the *CD* stage, a further increase in μ values takes place. This indicates that external mechanical stresses have led to a transition to the region of plastic strains with a change in the volume of the composite material.

4. Conclusion

The analysis of the obtained distribution patterns of strain fields, as well as deformation diagrams of steel concrete columns has shown that the compression of the concrete core with the steel shell prevents formation of microcracks and leads to an increase in the ultimate elastoplastic deformation. Since stresses in the concrete core exceed its prismatic strength, the value of the yield strength of a steel pipe can be taken as the value of critical stresses, taking into account elastoplastic deformation.

It has been established that elastic operation of the steel-concrete strut continues up to compressive stresses not exceeding 50 GPa. Further loading is accompanied by development of inelastic strain until stresses of 63 GPa are reached, corresponding to a loss of stability due to formation of a longitudinal bending of the rack.

The jump-like dependence of the Poisson's ratio on the longitudinal relative strain ε_{YY} of the steel-concrete column indicates redistribution of forces (deformations) among the components of the system at the stage of joint operation until separation of steel-concrete media and operation of concrete beyond elastic strain under volumetric compression.

References

[1] Afanasyev A and Kurochkin A 2016 Academia. Architecture and construction. 2 113–118.
[2] Duvanova I and Salmanov I 2016 Construction of unique buildings and structures. 6 89-103.
[3] Nishiyama I, Morino S, Sakino K and et al 2002 Summary of Research on Concrete-Filled Structural Steel Tube Column System Carried Out Under The US-JAPAN Cooperative Research Program on Composite and Hybrid Structures (The International Information Center for Structural Engineers, Japan)
[4] Belov N, Kopanitsa D, Yugov N and et al 2010 Mechanics of composite materials and structures. 16 181-190.
[5] Kumpyak O, Galyautdinov Z and Kokorin D 2016 AIP Conference Proceedings 1698 070006.
[6] Krishan A, Troshkina E and Astafieva M 2016 Bulletin of science and education of the North-West of Russia. 2(2) 1-7.
[7] Krishan Aand Remnev V. 2009 Industrial and Civil Engineering N10 22-24.
[8] Ustinov A, Kopanitsa D, Abzaev Yu, Klopotov A, Koshko B and Kopanitsa G 2017 AIP Conference Proceedings. 1800 040007.
[9] Sutton M A, Orteu Jean-Jose and Schreier H W 2009 Image Correlation for Shape, Motion and Deformation Measurements. Basic Concepts, Theory and Applications. (Springer Science, Business Media).
[10] Landau L D and Lifshnits E M 1965 The elasticity theory (Moscow, Nauka).

IOP Conf. Series: Materials Science and Engineering **696** (2019) 012021 doi:10.1088/1757-899X/696/1/012021

Turboexpander-refrigerant to reproduce arctic weather

A Yu Krainov and S M Gubanov

Tomsk State University, Tomsk, 634050 Russia

E-mail: akrainov@ftf.tsu.ru

Abstract. This paper describes a schematic diagram of an air cooler created on the basis of a turboexpander, presents its functioning principle and scopes of application in research and industry. The developed model of an air chiller allows cooling the air in the temperature range of minus 40÷140 °C. The air chiller has software control over time. The developed apparatus is proposed to use as a cooler to simulate the atmosphere of the Arctic in a research laboratory.

1. Introduction

Under the approved state program the planned amount of money allocated for the development of the Arctic until 2025 is more than 190 billion rubles. The second stage of the program (2018-2020) provides for the development and installation of a significant amount of equipment, mainly Russian-made in the Arctic zones. The third stage of the state program is aimed to create a leading scientific and technical reserve and technologies, including for the development of the shelf of the Arctic seas.

The conditions of the Arctic, where the temperature can reach minus 70 ° C, require a specific infrastructure. Materials, products, mechanisms created for the Arctic zone require special manufacturing. The design of the equipment demands a prerequisite engineering test in the chambers with temperatures up to minus 70 °C. Test methods should provide the ability to quickly change the temperature of the cooled samples according to a given law.

To provide the temperature tests the special climate chambers are used, which adapt multistage vapor compression machines as cooling devices. Steam compression machines have a high cost and low reliability. The steam compressors has the limit of the working temperature of minus 75 ° C. The machines have a complex control system, require highly qualified after-sales service, whilst requirements for test climate chambers are reliability, economical operation and simplicity of operation.

2. Results

This paper describes a successful actual experience with a unique new solution for “heat-cold” climate chambers. The cooling unit was created to automatically perform a thermostat control of several volumes at a temperature of minus 80°C. The unit equipped with the thermal regulator can vary the temperature according to a given law. It is possible to create temperature conditions in the range of cryogenic temperatures (up to minus 180°C).

The schematic diagram of the installation is shown in Fig. 1. The principle of adiabatic expansion of the air in a turboexpander device with the removal of the external work to an oil heat exchanger is used to generate cold in the chamber. Compressed air serves as the drive unit and the working medium of the refrigeration cycle.

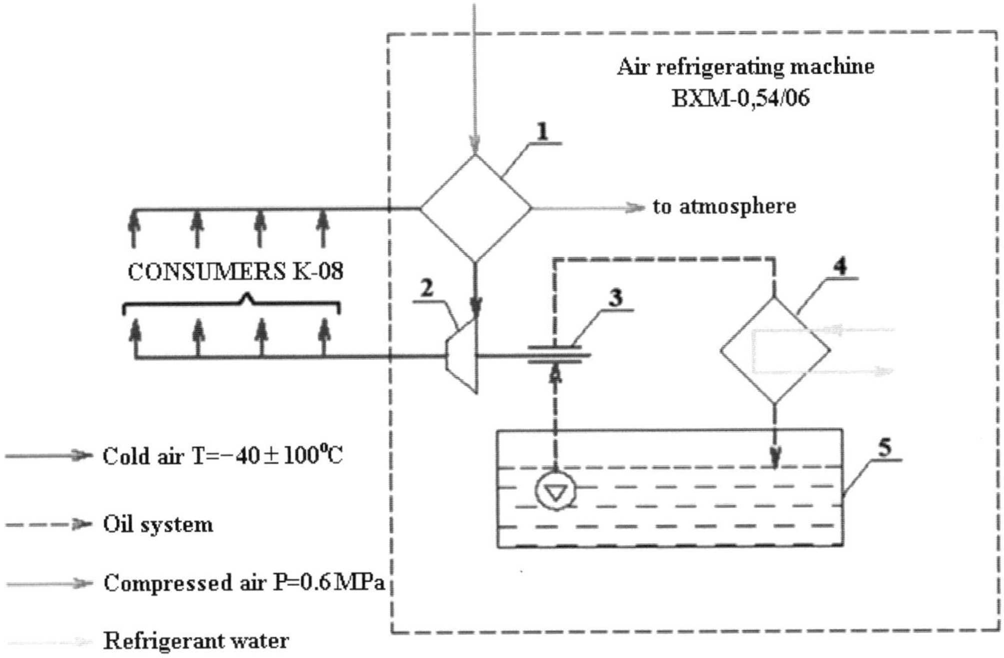

Figure 1. Schematic diagram of the air cooling device

Firstly the compressed air at the temperature of 20°C and dried to the dew point of minus 70°C enters a recuperative heat exchanger *1*, where it is cooled to the temperature of minus 40°C by the exhaust air from the consumer. Then the air comes into the turboexpander *2* and expands to the temperature of minus 100°C. Finally the air cools consumers and then the direct flow to heat exchanger *1*. The expansion work is removed by the oil brake *3*, where the oil is heated. The heated oil is cooled in the heat exchanger *4* and returns to the oil system *5*.

The cooling device can either be placed in a climate chamber or located at a distance from it. The cooling device is designed to transfer chilled air at a slight overpressure. This eliminates the use of heat exchangers and ventilation inside the chamber. The drive of the installation and the working medium is compressed air at a pressure of 0.6 MPa. It is possible to additionally use the exhaust air for conditioning and drying the room. The module is completely autonomous and equipped with a control system.

Six air refrigerating machines (BXM) were manufactured for the company of the Tomsk region JSC Siberian Chemical Combine. They have been in operation since 2014 and have shown high reliability and simplicity of operation.

The copyright of the BXM and the technologies used in the machine are protected by several Russian patents. The patent holder is Tomsk State University (TSU), LLC "APA-KANDT SIBERIA". LLC "APA-KANDT SIBERIA" and TSU concluded a partnership agreement on the use of intellectual property to manufacture the industrial designs.

Information on the BXM and its using is published in the papers [1-3]. We have registered the patents on the industrial design [4, 5]. During the introduction of the BXM, the related tasks have been solved. We have developed an air purifier for the turboexpander refrigeration machine and registered the patent [6]. Copyright certificates have been received for the turboexpander cooler control system and the calculation program of the cooling facility [7, 8]. We have also registered the patent on the cooling method using the BXM [9].

Information confirming the technical capabilities of the turboexpander cooler has been published in the scientific journal [3]. In addition to experiments with the target substance, the data on the cooling unit itself were processed. Five years of operating experience in a wide range of temperatures and loads showed additional advantages of the turbo-expander cooling method. We have also collected and analyzed data from 12 objects have being cooled by the BXM working on various "heating-cooling" algorithms.

It should be noted the flexibility of the BXM. The unique capabilities of the machine have been used to conduct a number of scientific researches. The studies have been carried out at cryogenic temperatures (up to minus 180 ° C). Experimental data have been obtained on the saturated anhydrous HF pressure dependence on temperature in the range from 77 K to 213 K. Based on the data we have formulated the equation of the pressure-temperature dependence for the saturated vapor HF, which is valid in the range from 140 K to 190 K [3]. We have also investigated the conditions of an effective removal process of HF from a gas medium in a dynamic mode in the presence of non-condensable gases.

Investigations on the gas-dynamic processes of the chilled air flow in the pipeline system and heat exchange with the walls of the tank make it possible to calculate the cooling characteristics of the given objects with high accuracy. Successful actual experience of the cooling project at cryogenic temperatures, as well as positive operating experience have shown the high advantages of the new cooling unit. In spite of the low refrigeration coefficient of the BXM at the temperature of minus 70°C (about 0.9) in comparison to the vapor compression units with a refrigeration coefficient of 1.5÷2, the first one has advantages. The maintenance costs are minimal, the unit provides savings in electrical energy due to the appropriate adjustment of the cold air temperature, the periodic maintenance of the BXM consists in heating the regenerative heat exchanger once a year during the routine maintenance equal to 4 hours.

3. Conclusions

Experience in solving various technical problems using the BXM-0,5/6 has shown the prospect of applying the unit as a cooler to simulate the atmosphere of the Arctic in a research laboratory or quality testing laboratory, and enterprise.

Acknowledgments

This study was financially supported by the Ministry of Education and Science of the Russian Federation as part of state assignment No. 9.9625.2017/8.9.

References:

[1] Gubanov S M, Krainov A Y , Mazur R L 2016 *Theoretical Foundations of Chemical Engineering* **50**, No 5 766-775
[2] Gubanov S M , Krainov A Y, Mazur R L, Durnovtsev M I 2017 *Theoretical Foundations of Chemical Engineering* **51** No 3 352-358
[3] Gubanov S M, Durnovtsev M I, Krainov A Yu 2017 *Journal of Engineering Physics and Thermophysics* **90** No 3 634–636
[4] Krainov A Y, Gubanov S M & The Tomsk State University, 2014, *Air refrigerating machine*, 2014108524
[5] Gubanov S M, Udut V M, Stulov V L & OJSC "Research and manufacturing combine "Heliymash" (NPO "Heliymash"), 2016, *Air refrigerating machine*, 2015501203.
[6] Gubanov S M, Krainov A Y & The Tomsk State University, 2015, *Air purifier for turboexpander refrigerator machine*, 2014108522.
[7] Gubanov S M, Krainov D A, Turboexpander Cooler Management Program, computer program, The Tomsk State University, Unified Register of Russian Programs for Electronic Computers and Databases, 2014611771
[8] Krainov A Y, Gubanov S M, Moiseeva K M, Durnovtsev M I, The program for calculating the cooling of the receiving tank set for gas desublimation. The Tomsk State University, Unified Register of Russian Programs for Electronic Computers and Databases, 2015614455
[9] Gubanov S M, Vasenin I M, Shrager E R, Krainov A Y, Zernaev P M, Mazur R L, Chukanov M V & JSC Siberian Chemical Combine, *The method of cooling the gas mixture*, 2013159280

IOP Conf. Series: Materials Science and Engineering **696** (2019) 012022 doi:10.1088/1757-899X/696/1/012022

Application of multilayer carbon nanotubes for creation of coatings absorbing electromagnetic radiation

A Shchegolkov[1], L Zavolokina[2], N Diesperova[2] and I Parfimovich[3]

[1]Institute of Technology, *Tambov State Technical University,* 106 Sovetskaya Street, Tambov 392000, Russian Federation
[2]Department of Applied Economics of the RUDN University, Miklukho-Maklaya Str. 6, Moscow 17198, Russian Federation
[3]Department of Elionics, A.N. Sevchenko Institute of Applied Physical Problems of Belarusian State University, 7, Kurchatova St., Minsk, 220045, Belarus

E-mail: Energynano@yandex.ru

Abstract. The article presents the use of carbon nanotubes as modifiers-materials for protection against electromagnetic radiation. The technique of obtaining nanomodified composites, as well as the method of characterization of carbon nanostructures is presented. Studies have shown that there is a linear dependence of the reflection coefficient on the frequency for composites modified multilayer carbon nanotubes based on Ni / 0.16 MgO catalyst. It was found that composites containing multilayer carbon nanotubes based on Ni / 0.16 MgO allow to achieve low values of the reflection coefficient and non-dependence of the reflection coefficient on the increase in frequency.

1. Introduction

The current state of scientific research in the field of radar absorbing coatings shows that an improvement in their characteristics can be achieved by creating new composite materials using nanoparticles [1]. In the past few years, much attention has been paid to functional materials with a hierarchical structure, since such an organization can allow the creation of new materials with improved operational characteristics [2].

Currently, there are hierarchical (multilayer) radar absorbing composite materials based on polymer matrices with good dielectric constant, but having a narrow range of operating frequencies. This disadvantage is due to the characteristics of the composite filler. Each filler or modifier has its own resonant frequency, at which the radio absorption will be most effective. The introduction of a large number of fillers to expand the operating range is impractical due to the increase in the cost of the resulting composite. To create a new type of composite material with reduced mass and size parameters, carbon nanostructured materials (fullerene, nanotubes, graphene, etc.) can be used as filler. This argument is partially confirmed in [1]. The development and creation of nanomodified materials is currently one of the relevant areas of scientific and applied research. Thanks to the modification, it is possible to create composite materials based on polymers that are resistant to mechanical stress [3], erosion wear [4], as well as materials with high electrical conductivity [5, 6].

[1] Corresponding author: Energynano@yandex.ru

Obtaining a new type of composite material with reduced mass and size parameters can be achieved using carbon nanostructures, in particular multilayer carbon nanotubes (MWCNTs). The effect of introducing a filler modifier, which is a MWCNT, into the polymer matrix, will depend on their morphological features. At the active sites of the catalyst, nucleation and growth of MWCNTs occur. Thus, the catalyst affects the morphology and structure of MWCNTs.

The aim of the work is to obtain a nanomodified composite using MWNTs and to study the frequency dependence of the transmittance of electromagnetic radiation.

2. Methods and Materials

The Ni / MgO catalyst system obtained by thermal decomposition was chosen as a catalyst for the synthesis of MWCNTs. The catalyst preparation method consisted in dissolving the starting components and thermal decomposition of the preliminary catalyst. The dissolution temperature of the starting components should not exceed 60 ° C, the thermal decomposition of the preliminary catalyst was carried out for 30 minutes at 500 ° C. The resulting catalyst was used for the synthesis of MWCNTs using CVD. As a carbon-containing raw material, a propane-butane mixture was used.

To assess the morphology and structure of MWCNTs synthesized on the obtained catalysts, transmission (TEM) and scanning electron microscopy (SEM) were used ("Hitachi H-800" (Hitachi, Japan)).

Polyurethane (Silagerm's two-component polyurethane compound) acts as a polymer matrix in which a hierarchical structure is created. The technique for introducing multilayer carbon nanotubes into polyurethane involves the following steps:

1) Adding weighted in accordance with the calculation of the MWCNTs in component A; the addition was carried out in small portions with constant stirring for 5 minutes;

2) After dispersion of the MWCNTs, component B was added to component B so that the volume ratio of component A (excluding MWCNTs) and component B was $\approx$ 1/1; then the composition is stirred for 10 minutes;

3) Further, when the polyurethane composite hardens, it is exposed to a magnetic field from neodymium magnets with a magnetic induction of 2000 mlT for 1 minute.

4) Polymerization of the nanocomposite in a vacuum cabinet at a temperature of 70 ° C for 5 hours.

To modify the polyurethane, a wt. 4% concentration was used. To determine the effects of the interaction of microwave electromagnetic radiation with composite materials (transmission, reflection, absorption of electromagnetic radiation), a scalar network analyzer based on a oscillating frequency generator, a rectangular waveguide measuring channel. The measurements were carried out in four frequency bands: 8–12 GHz, 12–18 GHz, 18–26 GHz, 26–40 GHz. The studied samples of the composites were cut in the form of a parallelepiped with dimensions of 23 × 10 × 2 mm, 16 × 8 × 2 mm, 11 × 5.5 × 2 mm, 7.2 × 3.4 × 2 mm.

3. Results and Discussion

The morphology of MWCNTs synthesized on Ni / 0.16MgO is shown in Fig. 1(a). The structure of the synthesized MWCNTs is shown in Figure 1 (b).

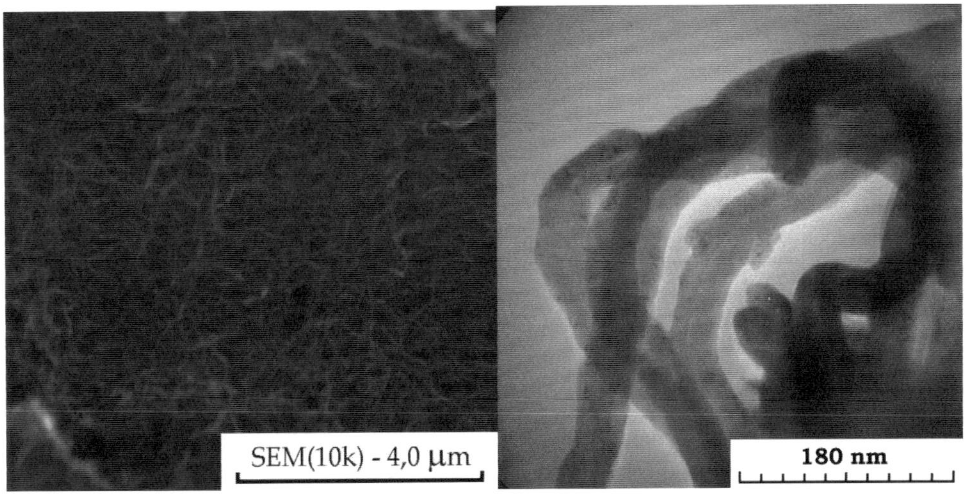

a) b)

Figure 1. a- SEM images MWCNTs (Ni / 0.16 MgO), b- TEM images MWCNTs (Ni / 0.16 MgO).

The diameter of MWCNTs synthesized on a Ni / 0.16 MgO catalyst ~ 30 ÷ 60 nm.
Figure 2 shows optical micrographs of a modified polyurethane with 4% MWCNTs.

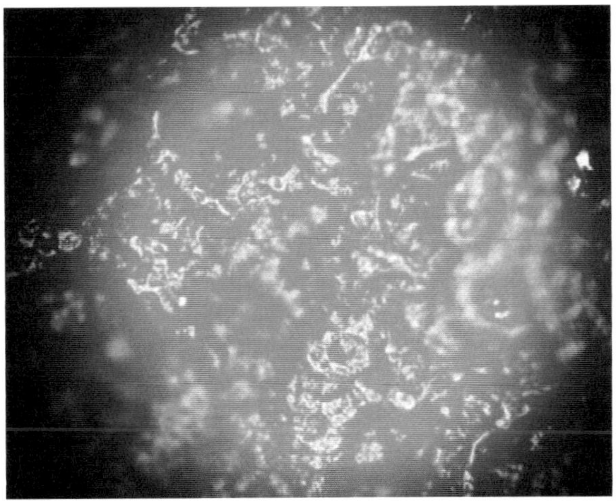

Figure 2. Optical micrographs of modified polyurethane with MWCNTs

From Figure 2 it follows that MWCNTs are fairly evenly distributed in the polymer matrix, which is the result of efficient distribution due to exposure to ultrasound.

The frequency dependence of the transmission of electromagnetic radiation by composite materials with MWNTs grown in the presence of a Ni / 0.16MgO catalyst is shown in Figure 3.

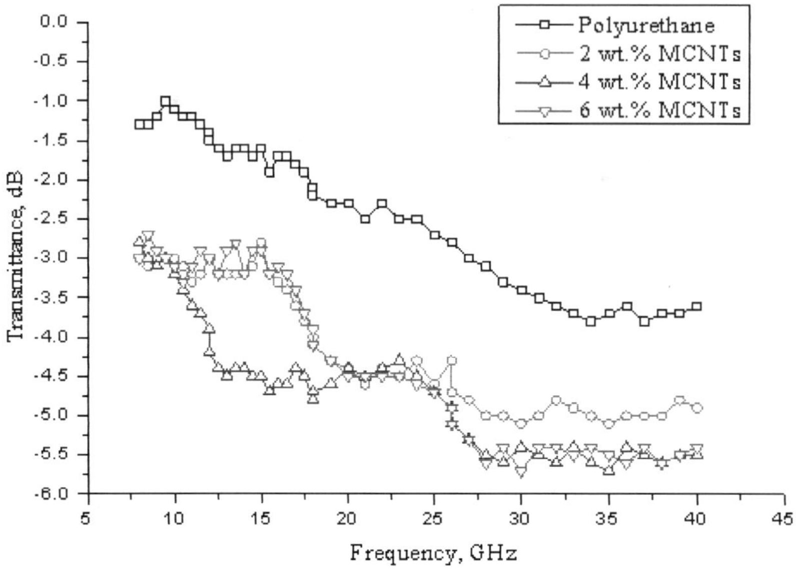

Figure 3. Frequency dependence of the transmittance of composite EMR materials with MWCNTs grown in the presence of a Ni / 0.16MgO catalyst.

For almost all samples of composite materials with MWCNTs grown on the Ni / 0.16MgO catalyst, similar frequency dependence and similar values of the EMR transmittance are observed — a decrease from 55% to 35% in the frequency range from 8 to 40 GHz. An exception is the sample with 4 wt.%, in which in the frequency range from 10 to 18 GHz there is a sharper decrease in the studied value in comparison with samples containing 2 and 6 wt.% MWCNTs.

The obtained results of the interaction of the nanomodified composite are related to the morphological structure of carbon nanotubes, namely their length, diameter, as well as the peculiarities of carbon nanotube interlacing in agglomerates. Changing the morphological features of carbon nanotubes can be used in the creation of layered coatings that effectively shield electromagnetic radiation.

4. Conclusion

The studies have shown that there is a linear dependence of the reflection coefficient on frequency for composites modified MWNTs Ni / 0.16MgO. For composites modified with MWNTs Ni / 0.16MgO, the reflection coefficient is practically independent of frequency and amounts to ~ 30–35% in the range from 8 to 40 GHz. The values of the transmission coefficients for the first and second systems monotonously decrease from 65% to 35% and from 55% to 35%, respectively, in the studied frequency range. Composites modified with MWNTs based on Ni / MgO 0.16 have a low value of the reflection coefficient. In this case, for these composites, there is an independence of the reflection coefficient from the increase in frequency.

Acknowledgments

This article is supported by the Program "5-100" of the Peoples' Friendship University of Russia.

References

[1] Zakharychev E *et al* 2016 *Bulletin Material Science* **39** 451

[2] Abrashova E, Gracheva I and Moshnikov V 2013 *Journal of Physics: Conference Series* **461** 012019
[3] Gao J, Li Ch, Shilpakar U, Shen Y 2015 *Materials and Design* **86** 289
[4] Jinhu C, Ian M, Tong D, Michael S and Krzysztof K 2014 *Carbon* **73** 421
[5] Mohammed H, Haya K and Yazan A 2013 *Composites: Part* A **46** 53
[6] Wolfgang B and Josef Z 2009 *Composites Science and Technolog* **69** 1486

IOP Conf. Series: Materials Science and Engineering **696** (2019) 012023 doi:10.1088/1757-899X/696/1/012023

Investigation of heating of nanomodified polymers under the action of microwave radiation

A Shchegolkov [1], A Semenov[2], M Kiryakova[2] and R Samofalova[2]

[1]Institute of Technology, *Tambov State Technical University,* 106 Sovetskaya Street, Tambov 392000, Russian Federation
[2]Department of Applied Economics of the RUDN University, Miklukho-Maklaya Str. 6, Moscow 17198, Russian Federation

E-mail: Energynano@yandex.ru

Abstract. The article presents the study of heating nanomodified polymers under the action of microwave radiation. Modification of polyethylene was carried out using CNTs synthesized on Co-Mo / Al2O3-MgO catalysts. The influence of CNTs concentration on the temperature regime of polymer heating is investigated. The study of interaction with microwave radiation was carried out at the emitter power of 800 W in 2 minute increments for 8 minutes

1. Introduction

The use of carbon nanotubes (CNTs) as polymer modifiers opens up new technological prospects in the creation of materials with functional properties [1-2]. CNTs can have dual functionality, i.e. provide microwave heating and improve electrical conductivity and mechanical properties. Thus, it is possible to develop polymer products with improved mechanical and electrical properties and functionality providing such possible technological operations as selective welding.

In this regard, it is important to study the influence of the nature of carbon materials on the interaction with microwave radiation in polymer matrices.

We studied the effect of the area of carbon particles on the interaction with microwave radiation. According to the results of studies it was concluded that the morphological characteristics of carbon had a strong influence on their properties under the influence of microwave heating. Studies carried out in [3] have shown that the efficiency of heating soot particles increases when the surface area of the particle is higher. In [4] the authors proposed a mechanism in which residual metals (Fe) of the catalyst for CNTs synthesis could influence the effect of microwave perception. However, other studies have shown the independence of the microwave effect from residual metals in CNTs [5].

In the article [6] polypropylene was chosen as a polymer matrix and nanocomposites were prepared by mixing in the melt using a twin-screw extruder. The efficiency of the heating process was tested taking into account the dispersion of CNTs in the polymer matrix, which allowed assessing the effect of dispersion as the main factor on heating.

In work [7] researches of dispersion of the polypropylene filled with 1% CNTs were carried out. Nanocomposites were obtained by compounding in the melt in a rotary twin-screw extruder. Processing parameters such as screw speed, screw configuration, and feed technology have been

[1] Corresponding author: Energynano@yandex.ru

modified to analyze their effect on carbon nanotube dispersion. According to the research, the question remains related to the uniformity of heat release and the influence of CNTs concentration on the heating dynamics of the nanomodified polymer.

However, the studies [3-7] do not fully reflect the influence of CNTs on the effects of interaction with microwave radiation, especially the influence of CNTs concentration in the polymer matrix.

The research objectives included:

1. Identification of the influence of CNTs concentration on the mode of polymer heating under the action of microwave.
2. Detection of exposure time under the action of microwave radiation to heat samples of different concentrations.

2. Methods and Materials

Polyethylene is used as a polymer matrix. Modification of polyethylene was carried out using CNTs synthesized on Co-Mo / Al2O3-MgO catalysts. The concentration of CNTs in the polymer reached 15%, other concentrations were wt.0.3 and 0.5 %. CNTs morphology was studied using scanning (SEM) and transmission electron microscope (TEM) “Hitachi H-800” (Hitachi, Japan). The Mooney polymer was mixed in a mixer. Further, the resulting material was extruded on a single-screw extruder with a drive motor power of 1.5 kW. The study of interaction with microwave radiation was carried out at the emitter power of 800 W with different time delay up to 8 minutes in increments of 2 minutes. The temperature field distribution on the composite surface was investigated using a Testo 875 thermal imager (Black Forest, Germany).

3. Results and Discussion

CNTs (Figure 1) have an outer diameter of about 25-35 nm and an inner diameter of 10-15 nm. CNTs having an outer diameter of 5-15 nm, have an inner diameter of 5-7 nm.

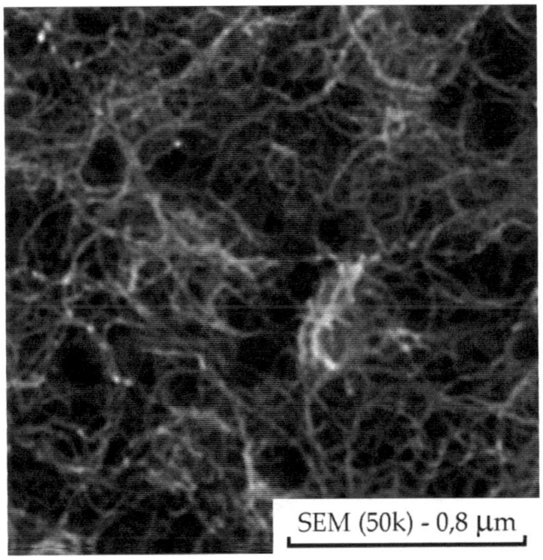

Figure 1. SEM CNTs

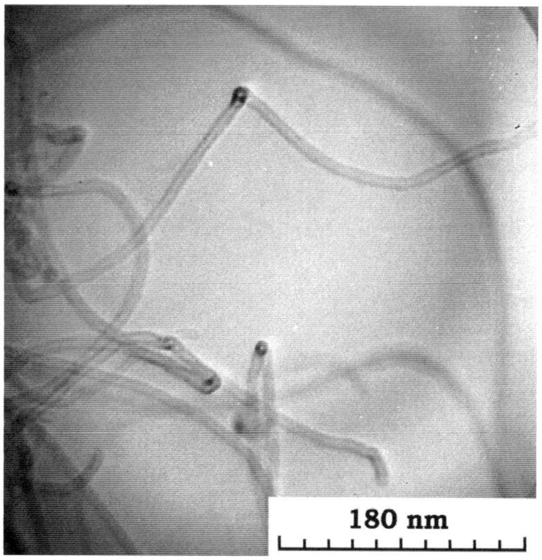

Figure 2. TEM CNTs

Nanomodified polymers obtained using an extruder, are shown in Figure 3.

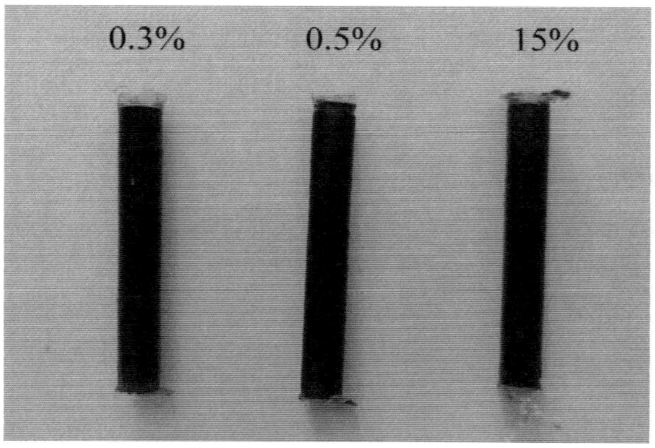

Figure.3. Nanomodified polymers.

Figures 4 - 6 show thermograms with the heat release of samples under the influence of microwave radiation.

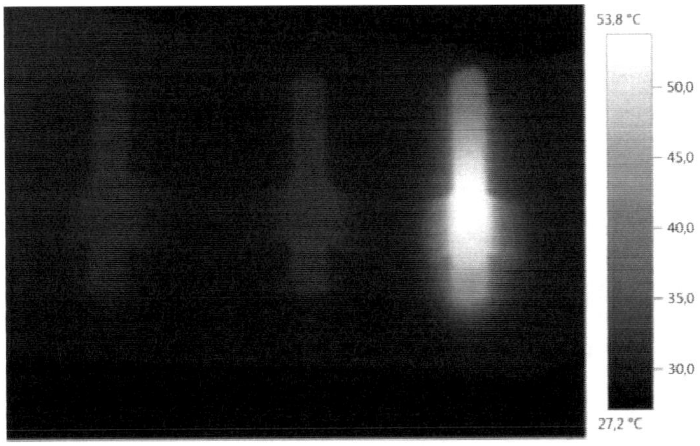

Figure 4. Thermogram of the samples (2 minutes)

The first experience showed (Figure 4) that the sample with the highest CNTs content was heated to 53.8 °C.

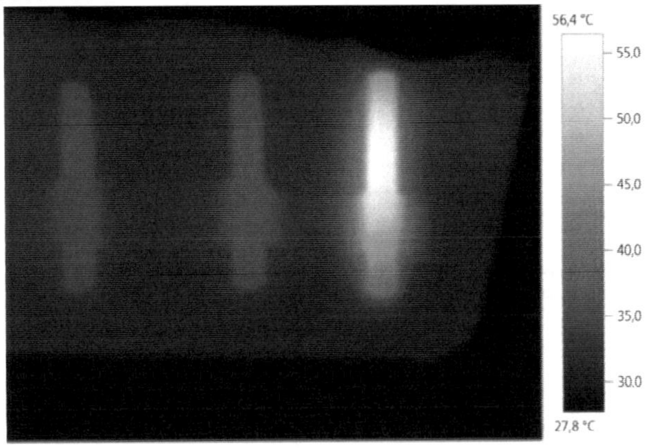

Figure 5. Thermogram of the samples (4 minutes)

In Figure 5 it is possible to observe an increase in temperature to 56.4 °C of the third sample containing 15% CNTs. However, there is also a slight heating of the first and second samples.

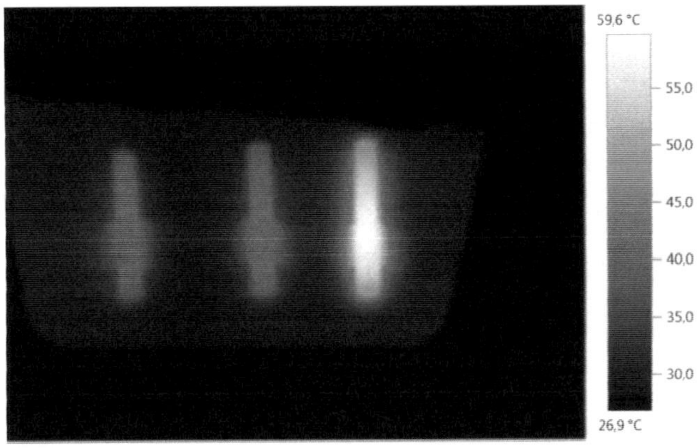

Figure 6. Thermogram of the samples (6 minutes)

The maximum temperature of the first and second samples is in the 40 °C zone (Figure 6).

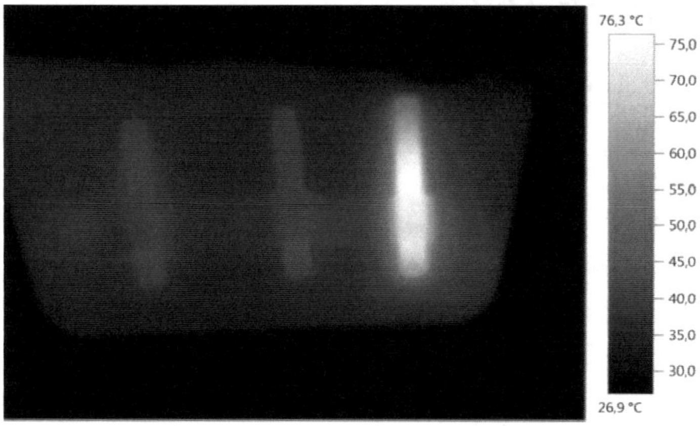

Figure 7. Thermogram of the samples (8 minutes)

The fourth stage of the experiment showed a large increase in temperature to 76.3 °C (Fig. 7). This is due to the fact that at this stage – the longest duration of the experiment 8 minutes. During this time, the samples gain the largest possible amount of thermal energy. At this concentration there is a peak of heat release in the Central region with a decrease in temperature to the ends of the sample.

Figure 8 shows the dynamics of temperature regime change for a sample with wt. 15 % CNTs.

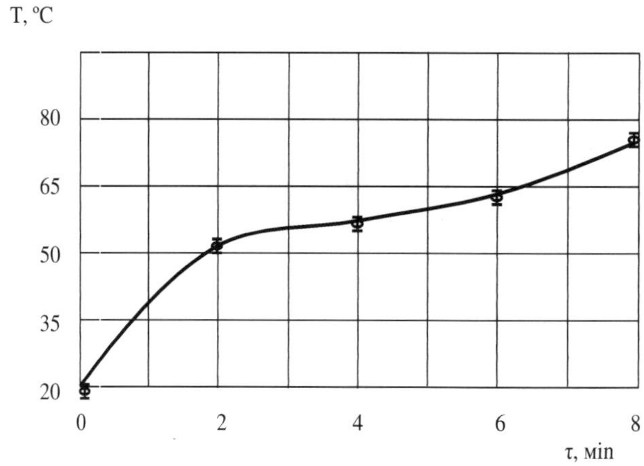

Figure 8. Dynamics of temperature change for the sample with wt. 15 % CNTs

The sharp temperature peak of 76.3 °C at 8 minutes can be explained by the fact that the heating occurred in the volume of the material over time, those CNTs that were located more deeply due to heat transfer gave their thermal energy to the upper layer of the polymer. Due to this, it is possible to explain the uneven temperature dynamics during microwave heating. This feature should be taken into account when long-term heating, as it is possible to ignite the polymer, due to excessive temperature increase of the inner layers.

4. Conclusion

Studies have shown how the concentration of CNTs in polymers affects the dynamics of heating depending on the time of exposure to microwave radiation. The highest temperature corresponded to the sample with a CNTs concentration of 15 % and was 76.3 °C. at this concentration, there is a peak of heat release in the Central region with a decrease in temperature to the ends of the sample.

Acknowledgments
This article is supported by the Program "5-100" of the Peoples' Friendship University of Russia.

References

[1] Soares B. G., Calheiros L. F., Silva A. A., Indrusiak T., Barra G. M., Livi S. 2018 *J. Appl.* Polymer Science, **135** 24
[2] J. Harper, D. Price, and J. Zhang, J. 2007 Microwave Power Electromagn. Energy, **40**, 219
[3] Q. Ling, J. Sun, Q. Zhao, and Q. Zhou 2009 Mater. Sci. Eng. B, 162
[4] A. Wadhawan, D. Garrett, and J.M. Perez 2003 Appl. Phys. Lett., **83** 2683
[5] F. Naab, M. Dhoubhadel, O.W. Holland, J.L. Duggan, J. Roberts, and F.D. McDaniel 2004 Proceedings Of the International Conference on PIXE and its Analytical Applications
[6] A. Benedito, I. Buezas, E. Gimenez, B. Galindo, and A. Ortega, J. 2011 Appl. Polym. Sci. **122** 3744
[7] Galindo, B., Benedito, A., Ramos, F., & Gimenez, E. 2016 Polymer Engineering & Science, **56** 12

Innovation policy in the Arctic as a part of the strategic approach of nation branding of Russia

Elena Danilova[1,2]
Financial University under the Government of the Russian Federation
49 Leningradsky Ave., Moscow, Russia
National Research Tomsk State University, 36 Lenin Ave., Tomsk, Russia

E-mail: elena.a.danilova@yandex.ru

Abstract. The paper presents the results of a research of the modern political interests and a role of Russia in the Arctic. Two main directions of development and presence of Russia in the Arctic region - economic (production of hydrocarbons) and military (military bases, the fleet, aircraft) are investigated. The conclusion is drawn on need of expansion of influence of Russia in the Arctic along with need of carrying out the balanced flexible policy for a prevention of the geopolitical conflict in the region.

1. Introduction
In the context of growing global competition, the position of innovative technologies as a part of implementing the strategic vision of nation branding has become a significant method of political marketing for detuning from global competitors based on exclusive competences. The foreign-policy situation additionally actualizes the correlation with nation branding in the category of prestige in world politics as a meaningful imaginary construct that provides global economic success [1, 2], which specifies superior armed forces [3]. The positioning of the government innovation policies in priority economic sectors for its significant regions is an effective element of place branding which is evaluated as a strategic tool of spatial planning and strategic communication structure to the foreign public [4, 5 ,6]. An important contemporary academic activity is the examination of the role of public administration for the development of place brands [7, 8, 9]. For the development of the strategic approach of nation branding, it is important to determine a symbolic influence on the strategic spatial areas of government policy. In this regard, the problem of the innovative exploration of the Arctic is highly relevant to contemporary Russian political discourse.

2. Results
At present, the Arctic region is on the focus of geopolitical interests for many of the world's powers, while not only five powers which geographically and historically claim the presence in the Arctic region (the current international law establishes the division of the Arctic into five sectors, along the Northern borders of Russia, the United States, Canada, Denmark (Greenland), and Norway), but also significantly remote from it, such as India, China, Japan. Due to the present-day geopolitical gap, the strengthening of military and economic influence in the Arctic is a task of decisive action for Russia.

Current political and economic opportunities of Russia in the struggle for the Arctic are largely based on the powerful scientific and technical groundwork of the Soviet period. The leadership of the USSR in the development of the North Pole is certainly beyond doubt: legendary Russian expeditions on icebreakers in the 1930s and the 1940s under the leadership of O. Schmidt, V. Wiese, et al; heroic epic of the rescue of the steamship Chelyuskin crew; launch of the first polar scientific drift station "North Pole-1" ("NP-1") under the command of I. Papanin; heroic transpolar nonstop flights across the Northern Pole by V. Chkalov; landing of four Soviet aircrafts at the Pole; other technical records and projects. All this provided insight into the territory where there is not only a place for a heroic deed but the area for the qualitative synthesis of national science, engineering, and technology [10]. However, due to the deep internal political crisis related to the Soviet Dissolution, the work of polar stations (NP) that provided a significant scientific and technical groundwork in the Arctic study during the Soviet period, was discontinued for a decade in the 1990s. While the other Arctic powers were actively carrying out scientific research and industrial development of the Arctic territory, in fact, Russia abandoned this process.

Since the early 2000s, with the advent of Vladimir Putin, political and expert circles began the active discussion about the need to regain Russian influence in the Arctic. In March 2000, as the Chairman of the Government and Acting President of the Russian Federation, Vladimir Putin held the All-Russian meeting in Murmansk on the development of the Russian Arctic sector, Eastern Siberia and the Russian Far East [11] that outlined the fundamental interest of new Russia in the Arctic. On 2008, the President of the Russian Federation Dmitry Medvedev approved the document "Principles of the state policy of the Russian Federation in the Arctic up to 2020 and beyond" [12]. Since 2014 the socio-economic development of the Arctic has been carried out within the framework of the state program of the Russian Federation that emphasizes the priority of the state interests in the region [13].

The President of the Russian Federation approved the strategy for the development of the Arctic Zone of the Russian Federation and provision of national security up to 2020. The priority areas for the exploration of the Arctic zone include: comprehensive socio-economic development of the region, development of science and technology, creation of modern information and telecommunications infrastructure, environmental security, international cooperation in the Arctic, military security, defence and protection of the state border of the Russian Federation in the Arctic [14]. The Strategy of the provision of national security of the Russian Federation highlights the growing role of the Arctic region in the implementation of Russian national interests in national security [15]. The new large-scale state armament programme, launched in 2018, gives particular attention to the adaptation of weapons and military equipment to the Arctic conditions [16]. The splash of political and scientific interest is also expressed in the regular convening of meetings on the Arctic exploration, in speeches of country's top leadership about the need to strengthen Russian presence in the Arctic region, extended coverage of the topic in mass media; in annual holding of the Days of the Arctic since 2010, and other applied research conferences. In 2015, the Arctic summit was held in Salekhard, and the international Arctic science, innovation, and exhibition campus EXPO-Arctic was opened. The information context contributes to the attention of the public and external players on the expansion of Russian presence in the Arctic that provides a solid foundation for applying this subject in the strategic approach of nation branding.

Arctic exploration has been performed by Russia in two main areas: expansion of the hydrocarbon production and strengthening of its military and political presence in the region. According to the Presidential Address to the Russian Federation Federal Assembly on December 2014, the Arctic is positioned as a priority direction for Russian economic policy. The Address emphasizes the need for a comprehensive project of contemporary competitive development of the North Sea Route as an effective transit route and a stimulator of business activity on the Russian Pacific coast in the exploration of Arctic territories [17]. Arctic gas and oil reserves present a potential major additional source of hydrocarbons for Russia. According to experts, the total reserves of fuel and energy resources of the Arctic part of Russian Federation exceed 1.6 trillion tons, and the continental shelf contains about a quarter of all offshore hydrocarbon reserves in the world [18].

The resource-based structure of the Russian economy and the unstable situation on the global geo-economic market determine Russia's strategic interest in the development of Arctic deposits. Under the conditions of the current political and economic situation, the two main subsoil users, Rosneft and Gazprom, found themselves in a complicated situation: these companies cannot attract foreign partners with their technologies to exploration and exploitation, and other Russian companies do not have access to the shelf due to legal requirement; at the same time Gazprom and Rosneft have to implement license commitments that force them to conduct the exploration according to the schedule of licenses, and require large financial investments. Rosneft owns 80% of the licensed areas on the Arctic shelf, 70% of which is a less explored Eastern part of the shelf. The company develops large poorly explored fields which also increases the need for investment in their development. At the same time, Gazprom does not seek the leadership by territories but it focuses on real field development in the near future. Gazprom has an operating Prirazlomnoye field on the Arctic shelf, and the Sakhalin shelf is also being developed. Gazprom, in turn, develops small fields which more explored and closer to existing fields. It reduces the relative need for investment in comparison with Rosneft's fields [19]. The largest innovative project of recent times can be considered the successful launch of the first and only one offshore ice-resistant fixed oil-producing platform "Prirazlomnay" owned by JSC "Gazprom Neft Shelf", from which the first commercial batch of oil was shipped in April 2014.

Due to insufficient knowledge of areas and extreme Arctic climatic conditions, the most advanced innovative technologies in the development of hydrocarbon fields are needed to overcome the dependence on foreign technologies and equipment. In March 2014 Deputy Prime Minister Dmitry Rogozin informed about the launch of the newest strategic project for developing of underwater (subglacial) technologies of autonomous (robotic) exploration of mineral deposits in the Arctic seas. Project developers face a serious technological challenge. It is proposed to set up "unmanned multimodule complexes with a full production cycle. These are completely underwater "cities" with their transport, energy supply, and communication lines. The main elements of these complexes will be underwater exploration vessels and carrier vessel, equipment for drilling, production and preparation of products, power supply, operation and repair, integrated security" [20]. As a part of the conversion, the project will involve the resources of the military-industrial complex. As a result, the industrial development of the Arctic will also serve to strengthen the military and political presence of Russia in the region. At the same time, the aggravation of the worldwide geopolitical situation and sanctions against Russia could not but impact on the political and economic situation in the Arctic. Practically all company's projects in difficult fields of the western waters of the Russian Arctic were associated with foreign partners: ExxonMobil, Statoil, Eni. Due to sanctions, ExxonMobil indefinitely discontinued cooperation with Rosneft on these fields. Representatives of the other two companies have noted that the projects are not discontinued but are subject to adjustment [19]. Therefore, Russia faces a serious technological challenge for the rational use of foreign investment for own innovative development work. In addition, today, the defense of territories of the subarctic shelf, which Russia claims, is likely to be more difficult.

In relation with the sharp deterioration of the geopolitical situation, the country's leadership deliberately focuses the attention of the population, world community, and other geopolitical players on the subject of expanding Russian military, political and economic presence in the Arctic. In April 2014, the President of the Russian Federation Vladimir Putin held a historically significant broadened meeting of the Security Council of the Russian Federation. It was announced an upcoming creation of a special government body for the implementation of the state policy in the Arctic on the strategic approaches of national security [18]. Thus, the President of the Russian Federation Vladimir Putin instructed to create an integrated system of basing surface ships and submarines of the new generation, support the development of ice-breaker fleet (in this regard, Russia is a world leader but the agenda still contains the problem of ice-breaker ageing and the need of new technological solutions on the subject). A milestone event can be considered the construction and transfer to the Russian Navy in December 2014 of the ballistic missile submarine of the fourth generation "Vladimir Monomakh", from the board of which in September 2014 the ballistic missile "Bulava" was successfully launched as a part of state test operations [21]. Such communication reasons should systematically fill the strategy of nation branding on an ongoing basis. In April 2019 during the annual international forum "Arctic: Territory of Dialogue", the President of the Russian Federation Vladimir Putin reiterated Russian intention to strengthen its military, political and economic presence in the Arctic, in particular, the embracing of a new strategy of the exploration of the Russian Arctic until 2035, the development of the Russian Arctic icebreaker fleet which by 2035 will consist of at least 13 heavy linear icebreakers, including 9 atomic ones [22].

Russian military presence in the Arctic is related to the political, economic and scientific interests. In September 2013 the Ministry of Defence of the Russian Federation resumed permanent Russian military presence in the Arctic, organizing a campaign of the surface combatant squadron of the Northern fleet led by the heavy nuclear missile cruiser "Pyotr Velikiy" in the sub-arctic regions of the Northern Sea Route [23]. In October 2013, the work of the military airfield "Temp" on Kotelny island was restored [24]. Today, the Joint Strategic Command "North" serves as a garrison. It is the only and most powerful formation of the Russian naval fleet, which has the status of a military district. After the creation of the "North" based on the Northern fleet on December 15, 2014, to date, it has increased its combat potential and showed its readiness to protect Russian national interests in the Arctic. The structure of the Joint Strategic Command "North" includes the union of atomic submarines, union of all-arms forces, air force and air defence, army corps of land forces, coastal defence troops, and naval surface force. The subunits of radio troops are deployed and carried out the tour of duty on a round-the-clock basis in the next archipelagos: Franz Josef Land, Novaya Zemlya, Severnaya Zemlya, Novosibirsk Islands, near the town of Tiksi, on Wrangel Island and Cape Schmidt. Military bases "Arctic Shamrock" and "Northern Clover" were put into operation [25]. The build-up of a systemic military presence contributes to the strengthening of Russia's geopolitical influence in the Arctic Region.

3. Conclusions

At the same time, it should be expected to take the parallel reciprocal steps from the Arctic countries of NATO that convert the Arctic into a zone of potential geopolitical conflict. Russian presence in the Arctic is necessary and possible, but in any case, only those states that will respond to the Arctic technological challenge with innovative solutions and development works will be able to claim political and economic influence in the region.

The expansion of Russian economic, military and political influence in the Arctic region has the potential to strengthen the position of the state in the international community. However, in current geopolitical realities, balanced state policy is extremely important to neutralize as much as possible a hotbed of tension in the North.

References

[1] Anholt S 2010 Places: identity, image and reputation (London: Palgrave Macmillan) 168 p.
[2] Papadopoulos N et al 2016 Nation branding for foreign direct investment: An integrative review and directions for research and strategy (Journal of Product and Brand Management **25 7** pp. 615-628.
[3] Savage J 2011 The stability and breakdown of empire: European informal empire in China, the Ottoman Empire and Egypt (European Journal of International Relations **17 2** pp. 161–185.
[4] Oliveira E 2015 Place branding as a strategic spatial planning instrument (Place Branding and Public Diplomacy **11** 1 pp 18–33.
[5] Hankinson G 2010 Place branding theory: A cross-domain literature review from a marketing perspective (Cheltenham, UK: Edward Elgar Publishing Towards Effective Place Brand Management-Branding European Cities and Regions) pp 15–35.
[6] Van Assche K et al 2011 Planning, preservation and place branding: A tale of sharing assets and narratives (Place Branding and Public Diplomacy **7 2** pp 116–126.
[7] Warnaby G et al 2015 Sketching Futures for Place Branding Rethinking Place Branding, *Comprehensive Brand Development for Cities and Regions* (Cham: Springer) pp 241–248.
[8] Eshuis J et al 2012 Branding in Governance and Public Management (London: Routledge) 188 p.
[9] Cerda-Bertomeu M J et al 2016 Stakeholders' perceptions of place branding and the role of the public sector: An exploratory analysis (Place Branding and Public Diplomacy **12** pp 299–313.
[10] Scherbinin A et al 2015 The Russian Arctic: innovative possibilities at the turn of the past and the future (IOP Conference Series: Earth and Environmental Science vol 27 No. 1) URL: http://iopscience.iop.org/article/10.1088/1755-1315/27/1/012022/pdf (access date: 27.09.2019).
[11] Order No. 96 of the Government of the Russian Federation of January 20, 2000. "Holding on March 24, 2000, in Murmansk the All-Russian meeting on the problem of the development of the Russian

sector of the Arctic, Eastern Siberia and the Russian Far East" 2000 (Legal information portal "Poisk zakona" January 20) URL: http://poisk-zakona.ru/169390.html (access date: 03.08.2019).
[12] Decree No. 1969 of the President of the Russian Federation of September 18, 2008. "Principles of the state policy of the Russian Federation in the Arctic up to 2020 and beyond" 2008 (The Government of the Russian Federation, September 18) URL: http://government.ru/info/18359/ (access date: 03.08.2019).
[13] Decision No. 366 of the Government of the Russian Federation of April 21, 2014 "On approval of the state programme of the Russian Federation – Socioeconomic Development of the Russian Arctic Zone up to 2020" 2014 (The Government of the Russian Federation April 21) URL: http://government.ru/media/files/41d4d600d9800a20c26f.pdf (access date: 13.09.2019).
[14] "The strategy for the development of the Arctic Zone of the Russian Federation and provision of national security up to 2020" 2013 (Government of the Russian Federation February 20) URL: http://government.ru/info/18360/ (access date: 11.09.2019).
[15] Decree No. 683 of the President of the Russian Federation of December 31, 2015. "The strategy of national security of the Russian Federation" 2015 (The President of the Russian Federation. December 31) URL: http://www.kremlin.ru/acts/bank/40391 (access date: 19.02.2018).
[16] State armaments programme of Russia. Dossier 2018 (TASS February 26) URL: https://tass.ru/info/4987920 (access date: 19.08.2019).
[17] Presidential Address to the Federal Assembly of December 4, 2014 2014 (The President of the Russian Federation December 4) URL: http://www.kremlin.ru/transcripts/47173 (access date: 13.09.2019).
[18] Meeting of the Security Council on the implementation of Russia's state policy in the Arctic 2014 (The President of the Russian Federation April 22) URL: http://state.kremlin.ru/face/20845 (access date: 03.09.2019).
[19] Buzovsky V V 2016 Factor analysis of Russian Arctic shelf exploration. Strategic differences between approaches of PJSC "NK Rosneft" and PJSC "Gazprom" URL: http://pro-arctic.ru/01/11/2016/resources/23925 (access date 13.09.2019)
[20] Rogozin D 2014 Let's look into the abyss (Rossiyskaya Gazeta March 14 No. **6331** (59) p 17.
[21] From the board of atomic submarine "Vladimir Monomakh", the launch of "Bulava" was successfully executed 2014 (JSC PO Sevmash September 11) URL: http://www.sevmash.ru/rus/news/1769--l-r-lr-.html (access date: 14.09.2019).
[22] Putin: a new development strategy for the Russian Arctic until 2035 will be adopted this year 2019 (TASS April 9). URL: https://tass.ru/ekonomika/6312429 (access date: 19.08.2019).
[23] The Ministry of Defence resumes Russian permanent military presence in the Arctic 2013 (RIA Novosti September 14). URL: https://ria.ru/20130914/963158271.html?in=t (access date: 21.08.2019).
[24] The Russian airfield "Temp" in the Arctic resumes its work 2013 (RIA Lenta.ru October 29) URL: https://lenta.ru/news/2013/10/29/temp/ (access date: 10.09.2019).
[25] "Arctic Shamrock" and "Northern Clover" 2018 (Armeyskiy sbornik April 25) URL: http://army.milportal.ru/arkticheskij-trilistnik-i-severnyj-klever/ (access date: 09.08.2019).

AUTHOR INDEX